Logic

FOR

DUMMIES®

by Mark Zegarelli

WILEY

John Wiley & Sons, Inc.

Logic For Dummies®

Published by
John Wiley & Sons, Inc.
111 River St.
Hoboken, NJ 07030-5774
www.wiley.com

For general information on our other products and services, please contact our Customer Care Department within the U.S. at 877-762-2974, outside the U.S. at 317-572-3993, or fax 317-572-4002.

For technical support, please visit www.wiley.com/techsupport.

Wiley also publishes its books in a variety of electronic formats. Some content that appears in print may not be available in electronic books.

Library of Congress Control Number: 2006934804

ISBN-13: 978-0-471-79941-2

Manufactured in the United States of America

SKY10029338_082421

1B/RT/RR/QW/IN

About the Author

Mark Zegarelli is a professional writer with degrees in both English and Math from Rutgers University. He has earned his living for many years writing vast quantities of logic puzzles, a hefty chunk of software documentation, and the occasional book or film review. Along the way, he's also paid a few bills doing housecleaning, decorative painting, and (for ten hours) retail sales. He likes writing best, though.

Mark lives mostly in Long Branch, New Jersey, and sporadically in San Francisco, California.

Dedication

This is for Mark Dembrowski, with love for his unfailing support, encouragement, and wisdom.

Author's Acknowledgments

Writers don't write, they rewrite — and rewriting sure is easier with a team of first-rate editors to help. Many thanks to Kathy Cox, Mike Baker, Darren Meiss, Elizabeth Rea, and Jessica Smith of Wiley Publications for their eagle-eyed guidance. You made this book possible.

I would like to thank Professor Kenneth Wolfe of St. John's University, Professor Darko Sarenac of Stanford University, and Professor David Nacin of William Paterson University for their invaluable technical reviewing, and to Professor Edward Haertel of Stanford University for his encouragement and assistance. You made this book better.

Thanks also for motivational support to Tami Zegarelli, Michael Konopko, David Feaster, Dr. Barbara Becker Holstein, the folks at Sunset Landing in Asbury Park, and Dolores Park Cafe in San Francisco, and the QBs. You made this book joyful.

Publisher's Acknowledgments

We're proud of this book; please send us your comments through our Dummies online registration form located at www.dummies.com/register/.

Some of the people who helped bring this book to market include the following:

Acquisitions, Editorial, and Media Development

Project Editors: Mike Baker, Darren Meiss

Acquisitions Editor: Lindsay Lefevere

Copy Editors: Jessica Smith, Elizabeth Rea

Editorial Program Coordinator: Hanna K. Scott

Technical Editor: Kenneth Wolfe

Editorial Managers: Carmen Krikorian

Media Development Manager: Laura VanWinkle

Editorial Assistant: Erin Calligan

Cartoons: Rich Tennant
(www.the5thwave.com)

Composition Services

Project Coordinator: Jennifer Theriot

Layout and Graphics: Claudia Bell, Peter Gaunt, Brooke Graczyk, Denny Hager, Stephanie D. Jumper, Heather Ryan

Anniversary Logo Design: Richard Pacifico

Proofreaders: John Greenough, Joanne Keaton

Indexer: Valene Hayes Perry

Publishing and Editorial for Consumer Dummies

Diane Graves Steele, Vice President and Publisher, Consumer Dummies

Joyce Pepple, Acquisitions Director, Consumer Dummies

Kristin A. Cocks, Product Development Director, Consumer Dummies

Michael Spring, Vice President and Publisher, Travel

Kelly Regan, Editorial Director, Travel

Publishing for Technology Dummies

Andy Cummings, Vice President and Publisher, Dummies Technology/General User

Composition Services

Gerry Fahey, Vice President of Production Services

Debbie Stailey, Director of Composition Services

Contents at a Glance

Table of Contents

Introduction

. . . .

*Y*ou use logic every day — and I bet you didn't even realize it. For instance, consider these examples of times when you might use logic:

✔ Planning an evening out with a friend

✔ Asking your boss for a day off or for a raise

✔ Picking out a shirt to buy among several that you like

✔ Explaining to your kids why homework comes before TV

At all of these times, you use logic to clarify your thinking and get other people to see things from your perspective.

Even if you don't always act on it, logic is natural — at least to humans. And logic is one of the big reasons why humans have lasted so long on a planet filled with lots of other creatures that are bigger, faster, more numerous, and more ferocious.

And because logic is already a part of your life, after you notice it, you'll see it working (or *not* working) everywhere you look.

This book is designed to show you how logic arises naturally in daily life. Once you see that, you can refine certain types of thinking down to their essence. Logic gives you the tools for working with what you already know (the premises) to get you to the next step (the conclusion). Logic is also great for helping you spot the flaws in arguments — unsoundness, hidden assumptions, or just plain unclear thinking.

About This Book

Logic has been around a long time — almost 2,400 years and counting! So, with so many people (past and present) thinking and writing about logic, you may find it difficult to know where to begin. But, never fear, I wrote this book with you in mind.

If you're taking an introductory course in logic, you can supplement your knowledge with this book. Just about everything you're studying in class is explained here simply, with lots of step-by-step examples. At the same time, if you're just interested in seeing what logic is all about, this book is also a great place for you to start.

Logic For Dummies is for anybody who wants to know about logic — what it is, where it came from, why it was invented, and even where it may be going. If you're taking a course in logic, you'll find the ideas that you're studying explained clearly, with lots of examples of the types of problems your professor will ask you to do. In this book, I give you an overview of logic in its many forms and provide you with a solid base of knowledge to build upon.

Logic is one of the few areas of study taught in two different college departments: math and philosophy. The reason that logic can fit into two seemingly different categories is historical: Logic was founded by Aristotle and developed by philosophers for centuries. But, about 150 years ago, mathematicians found that logic was an indispensable tool for grounding their work as it became more and more abstract.

One of the most important results of this overlap is formal logic, which takes ideas from philosophical logic and applies them in a mathematical framework. Formal logic is usually taught in philosophy departments as a purely computational (that is, mathematical) pursuit.

When writing this book, I tried to balance both of these aspects of logic. Generally speaking, the book begins where logic began — with philosophy — and ends where it has been taken — in mathematics.

Conventions Used in This Book

To help you navigate through this book, we use the following conventions:

- *Italics* are used for emphasis and to highlight new words and terms defined in the text. They're also used for variables in equations.

- **Boldfaced** text indicates keywords in bulleted lists and also true (**T**) and false (**F**) values in equations and tables. It's also used for the 18 rules of inference in SL and the 5 rules of inference in QL.

- Sidebars are shaded gray boxes that contain text that's interesting to know but not critical to your understanding of the chapter or topic.

- Twelve-point boldfaced text (**T** and **F**) text is used in running examples of truth tables and quick tables to indicate information that's just been added. It's used in completed truth tables and quick tables to indicate the truth value of the entire statement.

- Parentheses are used throughout statements, instead of a combination of parentheses, brackets, and braces. Here's an example:

$$\sim((P \lor Q) \to \sim R)$$

What You're Not to Read

I would be thrilled if you sat down and read this book from cover to cover, but let's face it: No one has that kind of time these days. How much of this book you read depends on how much logic you already know and how thoroughly you want to get into it.

Do, however, feel free to skip anything marked with a Technical Stuff icon. This info, although interesting, is usually pretty techie and very skippable. You can also bypass any sidebars you see. These little asides often provide some offbeat or historical info, but they aren't essential to the other material in the chapter.

Foolish Assumptions

Here are a few things we've assumed about you:

- ✔ You want to find out more about logic, whether you're taking a course or just curious.

- ✔ You can distinguish between true and false statements about commonly known facts, such as "George Washington was the first president," and "The Statue of Liberty is in Tallahassee."

- ✔ You understand simple math equations.

- ✔ You can grasp really simple algebra, such as solving for x in the equation $7 - x = 5$

How This Book Is Organized

This book is separated into six parts. Even though each part builds on the information from earlier parts, the book is still arranged in a modular way. So, feel free to skip around as you like. For example, when I discuss a new topic that depends on more basic material, I refer you to the chapter where I introduced those basics. If, for right now, you only need info on a certain topic, check out the index or the table of contents — they'll for sure point you in the right direction.

Here's a thumbnail sketch of what the book covers:

Part 1: Overview of Logic

What is logic? What does it mean to think logically, or for that matter illogically, and how can you tell? Part I answers these questions (and more!). The chapters in this part discuss the structure of a logical argument, explain what premises and conclusions are, and track the development of logic in its many forms, from the Greeks all the way to the Vulcans.

Part II: Formal Sentential Logic (SL)

Part II is your introduction to formal logic. Formal logic, also called symbolic logic, uses its own set of symbols to take the place of sentences in a natural language such as English. The great advantage of formal logic is that it's an easy and clear way to express logical statements that would be long and complicated in English (or Swahili).

You discover sentential logic (SL for short) and the five logical operators that make up this form. I also show how to translate back and forth between English and SL. Finally, I help you understand how to evaluate a statement to decide whether it's true or false using three simple tools: truth tables, quick tables, and truth trees.

Part III: Proofs, Syntax, and Semantics in SL

Just like any logic geek, I'm sure you're dying to know how to write proofs in SL — yeah, those pesky formal arguments that link a set of premises to a conclusion using the rules of inference. Well, you're in luck. In this part, you discover the ins and outs of proof writing. You also find out how to write conditional and indirect proofs, and how to attack proofs as efficiently as possible using a variety of proof strategies.

You also begin looking at SL from a wider perspective, examining it on the levels of both syntax and semantics.

You find out how to tell a statement from a string of symbols that just looks like a statement. I also discuss how the logical operators in SL allow you to build sentence functions that have one or more input values and an output value. From this perspective, you see how versatile SL is for expressing all possible sentence functions with a minimum of logical operators.

Part IV: Quantifier Logic (QL)

If you're looking to discover all there is to know about quantifier logic (or QL, for short), look no further: This part serves as your one-stop shopping introduction. QL encompasses everything from SL, but extends it in several important ways.

In this part, I show you how QL allows you to capture more intricacies of a statement in English by breaking it down into smaller parts than would be possible in SL. I also introduce the two quantification operators, which make it possible to express a wider variety of statements. Finally, I show you how to take what you already know about proofs and truth trees and put it to work in QL.

Part V: Modern Developments in Logic

The power and subtlety of logic becomes apparent as you examine the advances in this field over the last century. In this part, you see how logic made the 19th century dream of the computer a reality. I discuss how variations of post-classical logic, rooted in seemingly illogical assumptions, can be consistent and useful for describing real-world events.

I also show you how paradoxes fundamentally challenged logic at its very core. Paradoxes forced mathematicians to remove all ambiguities from logic by casting it in terms of axiom systems. Ultimately, paradoxes inspired one mathematician to harness paradox itself as a way to prove that logic has its limitations.

Part VI: The Part of Tens

Every *For Dummies* book contains a Part of Tens. Just for fun, this part of the book includes a few top-ten lists on a variety of topics: cool quotes, famous logicians, and pointers for passing exams.

Icons Used in This Book

Throughout this book, you'll find four icons that highlight different types of information:

I use this icon to point out the key ideas that you need to know. Make sure you understand the information in these paragraphs before reading on!

This icon highlights helpful hints that show you the easy way to get things done. Try them out, especially if you're enrolled in a logic course.

Don't skip these icons! They show you common errors that you want to avoid. The paragraphs that don this important icon help you recognize where these little traps are hiding so that you don't take a wrong step and fall in.

This icon alerts you to interesting, but unnecessary, trivia that you can read or skip over as you like.

Where to Go from Here

If you have some background in logic and you already have a handle on the Part I stuff, feel free to jump forward where the action is. Each part builds on the previous parts, so if you can read Part III with no problem, you probably don't need to concentrate on Parts I and II (unless of course you just want a little review).

If you're taking a logic course, you may want to read Parts III and IV carefully — you may even try to reproduce the proofs in those chapters with the book closed. Better to find out what you don't know while you're studying than while you're sweating out an exam!

If you're not taking a logic course — complete with a professor, exams, and a final grade — and you just want to discover the basics of logic, you may want to skip or simply skim the nitty-gritty examples of proofs in Parts III and IV. You'll still get a good sense of what logic is all about, but without the heavy lifting.

If you forge ahead to Parts IV and V, you're probably ready to tackle some fairly advanced ideas. If you're itching to get to some meaty logic, check out Chapter 22. This chapter on logical paradoxes has some really cool stuff to take your thinking to warp speed. Bon voyage!

Part I
Overview of Logic

In this part . . .

So, let me guess, you just started your first logic class and you're scrambling around trying to discover the ins and outs of logic as quickly as possible because you have your first test in 48 hours. Or, maybe you're not scrambling at all and you're just looking for some insight to boost your understanding. Either way, you've come to the right place.

In this part, you get a firsthand look at what logic is all about. Chapter 1 gives an overview of how you (whether you know it or not) use logic all the time to turn the facts that you know into a better understanding of the world. Chapter 2 presents the history of logic, with a look at the many types of logic that have been invented over the centuries. Finally, if you're itching to get started, flip to Chapter 3 for an explanation of the basic structure of a logical argument. Chapter 3 also focuses on key concepts such as premises and conclusions, and discusses how to test an argument for validity and soundness.

Chapter 1

What Is This Thing Called Logic?

*Y*ou and I live in an illogical world. If you doubt this fact, just flip on the evening news. Or really listen to the guy sitting at the next barstool. Or, better yet, spend the weekend with your in-laws.

With so many people thinking and acting illogically, why should you be any different? Wouldn't it be more sensible just to be as illogical as the rest of the human race?

Well, okay, being illogical on purpose is probably not the best idea. For one thing, how can it possibly be sensible to be illogical? For another, if you've picked this book up in the first place, you're probably not built to be illogical. Let's face it — some folks thrive on chaos (or claim to), while others don't.

In this chapter, I introduce you to the basics of logic and how it applies to your life. I tell you a few words and ideas that are key to logic. And, I touch very briefly on the connections between logic and math.

Getting a Logical Perspective

Whether you know it or not, you already understand a lot about logic. In fact, you already have a built-in logic detector. Don't believe me? Take this quick test to see whether you're logical:

Q: How many pancakes does it take to shingle a doghouse?

A: 23, because bananas don't have bones.

If the answer here seems illogical to you, that's a good sign that you're at least on your way to being logical. Why? Simply because if you can spot something that's illogical, you must have a decent sense of what actually is logical.

In this section, I start with what you *already* understand about logic (though you may not be aware of it), and build towards a foundation that will help you in your study of logic.

Bridging the gap from here to there

Most children are innately curious. They always want to know *why* everything is the way it is. And for every *because* they receive, they have one more *why*. For example, consider these common kid questions:

> Why does the sun rise in the morning?
>
> Why do I have to go to school?
>
> Why does the car start when you turn the key?
>
> Why do people break the law when they know they could go to jail?

When you think about it, there's a great mystery here: Even when the world doesn't make sense, why does it feel like it should?

Kids sense from an early age that even though they don't understand something, the answer must be somewhere. And they think, "If I'm here and the answer is there, what do I have to do to get there?" (Often, their answer is to bug their parents with more questions.)

Getting from here to there — from ignorance to understanding — is one of the main reasons logic came into existence. Logic grew out of an innate human need to make sense of the world and, as much as possible, gain some control over it.

Understanding cause and effect

One way to understand the world is to notice the connection between cause and effect.

As you grow from a child to an adult, you begin to piece together how one event causes another. Typically, these connections between cause and effect can be placed in an *if-statement*. For example, consider these if-statements:

If I let my favorite ball roll under the couch, *then* I can't reach it.

If I do all of my homework before Dad comes home, *then* he'll play catch with me before dinner.

If I practice on my own this summer, *then* in the fall the coach will pick me for the football team.

If I keep asking her out really nicely and with confidence, *then* eventually she will say yes.

Understanding how if-statements work is an important aspect of logic.

Breaking down if-statements

Every if-statement is made up of the following two smaller statements called *sub-statements*: The *antecedent,* which follows the word *if,* and the *consequent,* which follows the word *then.* For example, consider this if-statement:

If it is 5 p.m., *then* it's time to go home.

In this statement, the antecedent is the sub-statement

It is 5 p.m.

The consequent is the sub-statement

It's time to go home

Notice that these sub-statements stand as complete statements in their own right.

Stringing if-statements together

In many cases, the consequent of one if-statement becomes the antecedent of another. When this happens, you get a string of consequences, which the Greeks called a *sorites* (pronounced sore-it-tease). For example:

> If it's 5 p.m., then it's time to go home.
>
> If it's time to go home, then it's almost time for dinner.
>
> If it's almost time for dinner, then I need to call my husband to make reservations at the restaurant.

In this case, you can link these if-statements together to form a new if-statement:

If it's 5 p.m., then I need to call my husband to make reservations at the restaurant.

Thickening the plot

With more life experience, you may find that the connections between cause and effect become more and more sophisticated:

> *If* I let my favorite ball roll under the couch, *then* I can't reach it, *unless* I scream so loud that Grandma gets it for me, *though if* I do that more than once, *then* she gets annoyed and puts me back in my highchair.

> *If* I practice on my own this summer *but* not so hard that I blow my knees out, *then* in the fall the coach will pick me for the football team *only if* he has a position open, *but if* I do *not* practice, *then* the coach will *not* pick me.

Everything and more

As you begin to understand the world, you begin to make more general statements about it. For example:

> All horses are friendly.

> All boys are silly.

> Every teacher at that school is out to get me.

> Every time the phone rings, it's for my sister.

Words like *all* and *every* allow you to categorize things into *sets* (groups of objects) and *subsets* (groups within groups). For example, when you say "All horses are friendly," you mean that the set of all horses is *contained within* the set of all friendly things.

Existence itself

You also discover the world by figuring out what *exists* and *doesn't exist*. For example:

> Some of my teachers are nice.

> There is at least one girl in school who likes me.

> No one in the chess club can beat me.

> There is no such thing as a Martian.

Words like *some, there is,* and *there exists* show an overlapping of sets called an *intersection*. For example, when you say, "Some of my teachers are nice," you mean that there's an intersection between the set of your teachers and the set of nice things.

Similarly, words like *no, there is no,* and *none* show that there's no intersection between sets. For example, when you say "No one in the chess club can beat me," you mean that there's no intersection between the set of all the chess club members and the set of all the chess players who can beat you.

A few logical words

As you can see, certain words show up a lot as you begin to make logical connections. Some of these common words are:

if . . . then	and	but	or
not	unless	though	every
all	every	each	there is
there exists	some	there is no	none

Taking a closer look at words like these is an important job of logic because when you do this, you begin to see how these words allow you to divide the world in different ways (and therefore understand it better).

Building Logical Arguments

When people say "Let's be logical" about a given situation or problem, they usually mean "Let's follow these steps:"

1. Figure out what we know to be true.

2. Spend some time thinking about it.

3. Find the best course of action.

In logical terms, this three-step process involves building a *logical argument.* An argument contains a set of premises at the beginning and a conclusion at the end. In many cases, the premises and the conclusion will be linked by a series of intermediate steps. In the following sections, I discuss them in the order that you're likely to encounter them.

Generating premises

The *premises* are the facts of the matter: The statements that you know (or strongly believe) to be true. In many situations, writing down a set of premises is a great first step to problem solving.

For example, suppose you're a school board member trying to decide whether to endorse the construction of a new school that would open in September. Everyone is very excited about the project, but you make some phone calls and piece together your facts, or premises.

Premises:

> The funds for the project won't be available until March.
>
> The construction company won't begin work until they receive payment.
>
> The entire project will take at least eight months to complete.

So far, you only have a set of premises. But when you put them together, you're closer to the final product — your logical argument. In the next section, I show you how to combine the premises together.

Bridging the gap with intermediate steps

Sometimes an argument is just a set of premises followed by a conclusion. In many cases, however, an argument also includes *intermediate steps* that show how the premises lead incrementally to that conclusion.

Using the school construction example from the previous section, you may want to spell things out like this:

> According to the premises, we won't be able to pay the construction company until March, so they won't be done until at least eight months later, which is November. But, school begins in September. Therefore. . .

The word *therefore* indicates a conclusion and is the beginning of the final step, which I discuss in the next section.

Forming a conclusion

The *conclusion* is the outcome of your argument. If you've written the intermediate steps in a clear progression, the conclusion should be fairly obvious. For the school construction example I've been using, here it is:

Conclusion:

> The building won't be complete before school begins.

If the conclusion isn't obvious or doesn't make sense, something may be wrong with your argument. In some cases, an argument may not be valid. In others, you may have missing premises that you'll need to add.

Deciding whether the argument is valid

After you've built an argument, you need to be able to decide whether it's *valid,* which is to say it's a good argument.

To test an argument's validity, assume that all of the premises are true and then see whether the conclusion follows automatically from them. If the conclusion automatically follows, you know it's a valid argument. If not, the argument is *invalid*.

Understanding enthymemes

The school construction example argument may seem valid, but you also may have a few doubts. For example, if another source of funding became available, the construction company may start earlier and perhaps finish by September. Thus, the argument has a hidden premise called an *enthymeme* (pronounced en-thim-eem), as follows:

> There is no other source of funds for the project.

Logical arguments about real-world situations (in contrast to mathematical or scientific arguments) almost always have enthymemes. So, the clearer you become about the enthymemes hidden in an argument, the better chance you have of making sure your argument is valid.

Uncovering hidden premises in real-world arguments is more related to *rhetoric*, which is the study of how to make cogent and convincing arguments. I touch upon both rhetoric and other details about the structure of logical arguments in Chapter 3.

Making Logical Conclusions Simple with the Laws of Thought

As a basis for understanding logic, philosopher Bertrand Russell set down three laws of thought. These laws all have their basis in ideas dating back to Aristotle, who founded classical logic more than 2,300 years ago. (See Chapter 2 for more on the history of logic.)

All three laws are really basic and easy to understand. But, the important thing to note is that these laws allow you to make logical conclusions about statements even if you aren't familiar with the real-world circumstances that they're discussing.

The law of identity

The *law of identity* states that every individual thing is identical to itself.

For example:

> Susan Sarandon is Susan Sarandon.
>
> My cat, Ian, is my cat, Ian
>
> The Washington Monument is the Washington Monument.

Without any information about the world, you can see from logic alone that all of these statements are true. The law of identity tells you that any statement of the form "*X* is *X*," must be true. In other words, everything in the universe is the same as itself. In Chapter 19, you'll see how this law is explicitly applied to logic.

The law of the excluded middle

The *law of the excluded middle* states that every statement is either true or false.

For example, consider these two statements:

> My name is Mark.
>
> My name is Algernon.

Again, without any information about the world, you know logically that each of these statements is either true or false. By the law of the excluded middle, no third option is possible — in other words, statements can't be partially true or false. Rather, in logic, every statement is either completely true or completely false.

As it happens, I'm content that the first statement is true and relieved that the second is false.

The law of non-contradiction

The *law of non-contradiction* states that given a statement and its opposite, one is true and the other is false.

For example:

> My name is Algernon.
>
> My name is not Algernon.

Even if you didn't know my name, you could be sure from logic alone that one of these statements is true and the other is false. In other words, because of the law of contradiction, my name can't both be and not be Algernon.

Combining Logic and Math

Throughout this book, many times I prove my points with examples that use math. (Don't worry — there's nothing here that you didn't learn in fifth grade or before.) Math and logic go great together for two reasons, which I explain in the following sections.

Math is good for understanding logic

Throughout this book, as I'm explaining logic to you, I sometimes need examples that are clearly true or false to prove my points. As it turns out, math examples are great for this purpose because, in math, a statement is always either true or false, with no gray area between.

On the other hand, sometimes random facts about the world may be more subjective, or up for debate. For example, consider these two statements:

> George Washington was a great president.
>
> *Huckleberry Finn* is a lousy book.

Most people would probably agree in this case that the first statement is true and the second is false, but it's definitely up for debate. But, now look at these two statements:

> The number 7 is less than the number 8.
>
> Five is an even number.

Clearly, there's no disputing that the first statement is true and that the second is false.

Logic is good for understanding math

As I discuss earlier in this chapter, the laws of thought on which logic is based, such as the law of the excluded middle, depend on black-and-white thinking. And, well, nothing is more black and white than math. Even though you may find areas such as history, literature, politics, and the arts to be more fun, they contain many more shades of gray.

Math is built on logic as a house is built on a foundation. If you're interested in the connection between math and logic, check out Chapter 22, which focuses on how math starts with obvious facts called *axioms* and then uses logic to form interesting and complex conclusions called *theorems*.

Chapter 2

Logical Developments from Aristotle to the Computer

*W*hen you think about how *illogical* humans can be, it's surprising to discover how much logic has developed over the years. Here's just a partial list of some varieties of logic that are floating around out there in the big world of premises and conclusions:

Boolean logic	Modern logic	Quantifier logic
Classical logic	Multi-valued logic	Quantum logic
Formal logic	Non-classical logic	Sentential logic
Fuzzy logic	Predicate logic	Syllogistic logic
Informal logic	Propositional logic	Symbolic logic

As your eyes scan all of these varieties of logic, you may feel a sudden urge to embrace your humanity fully and leave logic to the Vulcans. The good news, as you'll soon discover, is that a lot of these varieties are quite similar. After you're familiar with a few of them, the rest become much easier to understand.

So, where did all of these types of logic come from? Well, that's a long story — in fact, it's a story that spans more than 2,000 years. I know 2,000 years seems like quite a lot to cram into one chapter, but don't worry because I guide you through only the most important details. So, get ready for your short history lesson.

Classical Logic — from Aristotle to the Enlightenment

The ancient Greeks had a hand in discovering just about everything, and logic is no exception. For example, Thales and Pythagoras applied logical argument to mathematics. Socrates and Plato applied similar types of reasoning to philosophical questions. But, the true founder of classical logic was Aristotle.

When I talk about *classical logic* in this section, I'm referring to the historical period in which logic was developed, in contrast with *modern logic*, which I discuss later in the chapter. But, classical logic can also mean the most standard type of logic (which most of this book is about) in contrast with *non-classical logic* (which I discuss in Chapter 21). I try to keep you clear as I go along.

Aristotle invents syllogistic logic

Before Aristotle (384–322 BC), logical argument was applied intuitively where appropriate in math, science, and philosophy. For example, given that all numbers are either even or odd, if you could show that a certain number wasn't even, you knew, then, that it must be odd. The Greeks excelled at this divide-and-conquer approach. They regularly used logic as a tool to examine the world.

Aristotle, however, was the first to recognize that the tool itself could be examined and developed. In six writings on logic — later assembled as a single work called *Organon,* which means *tool* — he analyzed how a logical argument functions. Aristotle hoped that logic, under his new formulation, would serve as a tool of thought that would help philosophers understand the world better.

Aristotle considered the goal of philosophy to be scientific knowledge, and saw the structure of scientific knowledge as logical. Using geometry as his model, he saw that science consisted of proofs, proofs of syllogisms, syllogisms of statements, and statements of terms. So, in the *Organon,* he worked from the bottom upwards: The first book, the *Categories,* deals with terms; the second, *On Interpretation,* with statements; the third, *Prior Analytics,* with syllogisms; and the fourth, *Posterior Analytics,* with proofs.

Prior Analytics, the third book in the *Organon* series, delves directly into what Aristotle called *syllogisms,* which are argument structures that, by their very design, appear to be indisputably valid.

The idea behind the syllogism was simple — so simple, in fact, that it had been taken for granted by philosophers and mathematicians until Aristotle noticed it. In a syllogism, the premises and conclusions fit together in such a way that, once you accept the premises as true, you must accept that the conclusion is true as well — regardless of the content of the actual argument being made.

For example, here's Aristotle's most famous syllogism:

Premises:

> All men are mortal.
>
> Socrates is a man.

Conclusion:

> Socrates is mortal.

The following argument is similar in form to the first. And it's the form of the argument, not the content, that makes it indisputable. Once you accept the premises as true, the conclusion follows as equally true.

Premises:

> All clowns are scary.
>
> Bobo is a clown.

Conclusion:

> Bobo is scary.

Categorizing categorical statements

Much of Aristotle's attention focused on understanding what he called *categorical statements*. Categorical statements are simply statements that talk about whole categories of objects or people. Furniture, chairs, birds, trees, red things, Meg Ryan movies, and cities that begin with the letter *T* are all examples of categories.

In keeping with the law of the excluded middle (which I discuss in Chapter 1), everything is either in a particular category or not in it. For example, a red chair is in the category of furniture, chairs, and red things, but not in the category of birds, trees, Meg Ryan movies, or cities that begin with the letter *T*.

Aristotle broke categorical statements down into the following two types :

> ✔ **Universal statements:** These are statements that tell you something about an entire category. Here's an example of a universal statement:
>
>> All dogs are loyal.

This statement relates two categories and tells you that everything in the category of dogs is also in the category of loyal things. You can consider this a universal statement because it tells you that loyalty is a universal quality of dogs.

✔ **Particular statements:** These are statements that tell you about the existence of at least one example within a category. Here's an example of a particular statement:

Some bears are dangerous.

This statement tells you that at least one item in the category of bears is also in the category of dangerous things. This statement is considered a particular statement because it tells you that at least one particular bear is dangerous.

Understanding the square of oppositions

The *square of oppositions* — a tool Aristotle developed for studying categorical statements — organizes the four basic forms of categorical statements that appear frequently in syllogisms. These four forms are based on the positive and negative forms of universal and particular statements.

Aristotle organized these four types of statements into a simple chart similar to Table 2-1. Aristotle's most famous example was based on the statement "All humans are mortal." However, the example in the table is inspired by my sleeping cat.

Table 2-1:	The Square of Oppositions	
	Positive Forms	*Negative Forms*
Universal Forms	**A:** *All cats are sleeping.*	**E:** *No cats are sleeping.*
	There doesn't exist a cat that isn't sleeping.	All cats are not sleeping.
	No cats are not sleeping.	There isn't a cat that is sleeping.
	Every cat is sleeping.	There doesn't exist a sleeping cat.
Particular Forms	**I:** Some cats are sleeping.	**O:** Not all cats are sleeping.
	Not all cats are not sleeping.	Some cats are not sleeping.
	At least one cat is sleeping.	There is at least one cat that isn't sleeping.
	There exists a sleeping cat.	Not every cat is sleeping.

As you can see from the table, each type of statement expresses a different relationship between the category of cats and the category of sleeping things. In English, you can express each type of statement in a variety of ways. I've listed a few of these in the table, but many more are possible in each case.

Aristotle noticed relationships among all of these types of statements. The most important of these relationships is the *contradictory* relationship between those statements that are diagonal from each other. With contradictory pairs, one statement is true and the other false.

For example, look at the **A** and **O** statements in Table 2-1. Clearly, if every cat in the world is sleeping at the moment, then **A** is true and **O** is false; otherwise, the situation is reversed. Similarly, look at the **E** and **I** statements. If every cat in the world is awake, then **E** is true and **I** is false; otherwise, the situation is reversed.

If you're wondering, the letters for the positive forms **A** and **I** are reputed to come from the Latin word *AffIrmo*, which means "I affirm." Similarly, the letters for the negative forms **E** and **O** are said to come from the Latin word *nEgO*, which means "I deny." The source of these designations is unclear, but you can rule out Aristotle, who spoke Greek, not Latin.

Euclid's axioms and theorems

Although Euclid (c. 325–265 BC) wasn't strictly a logician, his contributions to logic were undeniable.

Euclid is best known for his work in geometry, which is still called *Euclidean geometry* in his honor. His greatest achievement here was his logical organization of geometric principles into *axioms* and *theorems*.

Euclid began with five axioms (also called *postulates*) — true statements that he believed were simple and self-evident. From these axioms, he used logic to prove *theorems* — true statements that were more complex and not immediately obvious. In this way, he succeeded in proving the vast body of geometry logically followed from the five axioms alone. Mathematicians still use this logical organization of statements into axioms and theorems. For more on this topic, see Chapter 22.

Euclid also used a logical method called *indirect proof*. In this method, you assume the opposite of what you want to prove and then show that this assumption leads to a conclusion that's obviously incorrect.

For example, a detective in a murder mystery might reason: "If the butler committed the murder, then he must have been in the house between 7 p.m. and 8 p.m. But, witnesses saw him in the city twenty miles away during those hours, so he couldn't have also been in the house. Therefore, the butler didn't commit the murder."

Indirect proof is also called *proof by contradiction* and *reductio ad absurdum*, which is Latin for *reduced to an absurdity*. Flip to Chapter 11 for more about how to use indirect proof.

Chrysippus and the Stoics

While Aristotle's successors developed his work on the syllogistic logic of categorical statements, another Greek school of philosophy, the Stoics, took a different approach. They focused on *conditional statements*, which are statements that take the form *if . . . then. . . .* For example:

> *If* clouds are gathering in the west, *then* it will rain.

Most notable among these logicians was Chrysippus (279–206 BC). He examined arguments using statements that were in this if . . . then . . . form. For example:

Premises:

> If clouds are gathering in the west, then it will rain.
>
> Clouds are gathering in the west.

Conclusion:

> It will rain.

Certainly, there are connections between the Aristotelian and the Stoic approaches. Both focused on sets of premises containing statements that, when true, tended to fit together in a way that forced the conclusion to be true as well. But friction between the two schools of thought caused logic to develop in two separate strands for over a century, though over time these merged into a unified discipline.

Logic takes a vacation

After the Greeks, logic went on a very long vacation, with a few sporadic revivals.

Throughout the Roman Empire and Medieval Europe for over a thousand years, logic was often disregarded. Aristotle's writing on logic was occasionally translated with some commentary by the translator. However, very few people wrote original treatises on logic.

During the first millennium AD, more work on logic was done in the Arab world. Both Christian and Muslim philosophers in Baghdad continued to translate and comment upon Aristotle. Avicenna (980–1037) broke from this practice, studying logical concepts that involve time, such as *always*, *sometimes*, and *never*.

The 12th century saw a renewed interest in logic, especially *logical fallacies*, which are flaws in arguments. Some of this work, which was begun with Aristotle in his *Sophistic Refutations*, was used by theologians during this time of ever-widening Catholic influence in Europe. Throughout the next few centuries, philosophers continued to study questions of language and argument as they related to logic.

Additionally, as one of the seven liberal arts, logic was also central to the curriculum of the universities developing at this time. (I'm sure you are dying to know that the other six liberal arts were: grammar, rhetoric, arithmetic, geometry, astronomy, and music.)

Modern Logic — the 17th, 18th, and 19th Centuries

In Europe, as the Age of Faith gradually gave way to the Age of Reason in the 16th and 17th centuries, thinkers became optimistic about finding answers to questions about the nature of the universe.

Even though scientists (such as Isaac Newton) and philosophers (such as René Descartes) continued to believe in God, they looked beyond church teachings to provide answers as to how God's universe operated. Instead, they found that many of the mysteries of the world — such as the fall of an apple or the motion of the moon in space — could be explained mechanistically and predicted using mathematics. With this surge in scientific thought, logic rose to preeminence as a fundamental tool of reason.

Leibniz and the Renaissance

Gottfried Leibniz (1646–1716) was the greatest logician of the Renaissance in Europe. Like Aristotle, Leibniz saw the potential for logic to become an indispensable tool for understanding the world. He was the first logician to take

Aristotle's work a significant step farther by turning logical statements into symbols that could then be manipulated like numbers and equations. The result was the first crude attempt at *symbolic logic*.

In this way, Leibniz hoped logic would transform philosophy, politics, and even religion into pure calculation, providing a reliable method to answer all of life's mysteries with objectivity. In a famous quote from *The Art of Discovery* (1685), he says:

> The only way to rectify our reasonings is to make them as tangible as those of the Mathematicians, so that we can find our error at a glance, and when there are disputes among persons, we can simply say: "Let us calculate, without further ado, to see who is right."

Unfortunately, his dream of transforming all areas of life into calculation was not pursued by the generation that followed him. His ideas were so far ahead of their time that they were not recognized as important. After his death, Leibniz's writings on logic as symbolic calculation gathered dust for almost 200 years. By the time they were rediscovered, logicians had already caught up with these ideas and surpassed them. As a result, Leibniz was not nearly as influential as he might have been during this crucial phase in the development of logic.

Working up to formal logic

For the most part, logic was studied *informally* — that is, without the use of symbols in place of words — into the beginning of 19th century. Beginning with Leibniz, mathematicians and philosophers up to this time had improvised a wide variety of notations for common logical concepts. These systems, however, generally lacked any method for full-scale computation and calculation.

By the end of the 19th century, however, mathematicians had developed *formal logic* — also called *symbolic logic* — in which computable symbols stand for words and statements. Three key contributors to formal logic were George Boole, Georg Cantor, and Gottlob Frege.

Boolean algebra

Named for its inventor, George Boole (1815–1864), Boolean algebra is the first fully fleshed-out system that handles logic as calculation. For this reason, it's considered the precursor to formal logic.

Boolean algebra is similar to standard arithmetic in that it uses both numerical values and the familiar operations for addition and multiplication. Unlike arithmetic, however, only two numbers are used: 0 and 1, which signify *false* and *true*, respectively.

For example:

> Let A = Thomas Jefferson wrote the Declaration of Independence.
>
> Let B = Paris Hilton wrote the U.S. Constitution.

Because the first statement is true and the second is false (thank goodness!), you can say:

> $A = 1$ and $B = 0$

In Boolean algebra, addition is interpreted as *or*, so the statement

> Thomas Jefferson wrote the Declaration of Independence *or* Paris Hilton wrote the U.S. Constitution

is translated as

> $A + B = 1 + 0 = 1$

Because the Boolean equation evaluates to 1, the statement is true. Similarly, multiplication is interpreted as *and*, so the statement

> Thomas Jefferson wrote the Declaration of Independence *and* Paris Hilton wrote the U.S. Constitution

is translated as

> $A \times B = 1 \times 0 = 0$

In this case, the Boolean equation evaluates to 0, so the statement is false.

As you can see, the calculation of values is remarkably similar to arithmetic. But the meaning behind the numbers is pure logic.

Check out Chapter 14 for more on Boolean algebra.

Cantor's set theory

Set theory, pioneered by Georg Cantor in the 1870s, was another foreshadowing of formal logic, but with far wider influence and usefulness than Boolean algebra.

Loosely defined, a *set* is just a collection of things, which may or may not have something in common. Here are a few examples:

> {1, 2, 3, 4, 5, 6}
>
> {Batman, Wonder Woman, Spiderman}
>
> {Africa, Kelly Clarkson, November, Snoopy}

This simple construction is tremendously effective for characterizing important core ideas of logic. For example, consider this statement:

All U.S. states that contain the letter *z* begin with the letter *A*.

This statement can be verified by identifying the sets of all states that contain the letter *z* and begin with the letter *A*. Here are the two sets:

Set 1: {Arizona}   **Set 2:** {Alabama, Alaska, Arizona, Arkansas}

As you can see, every member of the first set is also a member of the second set. Thus, the first set is a *subset* of the second set, so the original statement is true.

Despite its apparent simplicity — or, rather, because of it — set theory would soon become the foundation of logic and, ultimately, of formal mathematics itself.

Frege's formal logic

Gottlob Frege (1848–1925) invented the first real system of formal logic. The system he invented is really one logical system embedded within another. The smaller system, *sentential logic* — also known as *propositional logic* — uses letters to stand for simple statements, which are then linked together using symbols for five key concepts: *not, and, or, if,* and *if and only if.* For example:

Let *E* = Evelyn is at the movies.

Let *P* = Peter is at home.

These definitions allow you to take these two statements:

Evelyn is at the movies *and* Peter is at home.

If Evelyn is at the movies, *then* Peter is *not* at home.

and turn them into symbols as follows:

E & *P*

E → ~*P*

In the first statement, the symbol & means *and*. In the second, the symbol → means *if. . .then,* and the symbol ~ means *not*.

I discuss sentential logic in more detail in Parts II and III.

The larger system, *quantifier logic* — also known as *predicate logic* — includes all of the rules from sentential logic, but expands upon them. Quantifier logic uses different letters to stand for the subject and the predicate of a simple statement. For example:

Let e = Evelyn

Let p = Peter

Let Mx = x is at the movies

Let Hx = x is at home

These definitions allow you to represent these two statements:

Evelyn is at the movies *and* Peter is at home

If Evelyn is at the movies, *then* Peter is *not* at home

as

Me & Hp

$Me \rightarrow \sim Hp$

Quantifier logic also includes two additional symbols for *all* and *some*, which allows you to represent the statements

Everyone is at the movies

Someone is at home

as

$\forall x \, [Mx]$

$\exists x \, [Hx]$

Quantifier logic has the power to represent the four basic categorical statements from Aristotle's square of oppositions (see the section "Categorizing categorical statements" earlier in the chapter). In fact, in its most developed form, quantifier logic is as powerful as all previous formulations of logic.

Check out Part IV for more on quantifier logic.

Logic in the 20th Century and Beyond

By the end of the 19th century, following Euclid's example (see "Euclid's axioms and theorems" earlier in this chapter), mathematicians sought to reduce everything in mathematics to a set of theorems logically dependent on a small number of axioms.

Frege, the inventor of the first real system of formal logic, saw the possibility that mathematics itself could be derived from logic and set theory. Beginning with only a few axioms about sets, he showed that numbers and, ultimately, all of mathematics followed logically from these axioms.

Frege's theory seemed to work well until Bertrand Russell (1872–1970) found a *paradox,* or an inconsistency that occurs when a set may contain itself as a member. Within Frege's system, Russell saw that it was possible to create a set containing every set that *doesn't* contain itself as a member. The problem here is that if this set contains itself, then it doesn't contain itself, and vice versa. This inconsistency became known as Russell's Paradox. (I discuss Russell's Paradox further in Chapter 22.)

Frege was devastated by the error, but Russell saw merit in his work. From 1910 to 1913, Bertrand Russell and Alfred North Whitehead produced the three-volume *Principia Mathematica,* a reworking of Frege's ideas grounding mathematics in axioms of set theory and logic.

Non-classical logic

The project of reducing math and logic to a short list of axioms opens up an interesting question: What happens if you start with a different set of axioms?

One possibility, for example, is to allow a statement to be something other than either true or false. In other words, you can allow a statement to violate the law of the excluded middle (see Chapter 1). The blatant violation of this law would have been unthinkable to the Greeks, but with logic formulated simply as a set of axioms, the possibility became available.

In 1917, Jan Lukasiewicz pioneered the first *multi-valued logic,* in which a statement may be not only *true* or *false,* but also *possible.* This system would be useful for classifying a statement such as this:

In the year 2162, the Yankees will win the World Series.

The introduction of category *possible* to the *true* and *false* pantheon was the first radical departure from *classical logic* — all of the logic that had existed before — into a new area called *non-classical logic.* (You can find out more about forms of non-classical logic — including fuzzy logic and quantum logic — in Chapter 21.)

Gödel's proof

The *Principia Mathematica,* the three-volume series written by Bertrand Russell and Alfred North Whitehead, firmly established logic as an essential foundation of math. However, more surprises were in store for logic.

With mathematics defined in terms of a set of axioms the question arose as to whether this new system was both *consistent* and *complete*. That is, was it possible to use logic to derive every true statement about math from these axioms, and no false statements?

In 1931, Kurt Gödel showed that an infinite number of mathematical statements are true but can't be proven given the axioms of the *Principia*. He also showed that any attempt to reduce math to a consistent system of axioms produces the same result: an infinite number of mathematical truths, called *undecidable statements*, which aren't provable within that system.

This result, called the *Incompleteness Theorem,* established Gödel as one of the greatest mathematicians of the 20th century.

In a sense, Gödel's Incompleteness Theorem provided a response to Leibniz's hope that logic would someday provide a method for calculating answers to all of life's mysteries. The response, unfortunately, was a definitive "No!" Logic — at least in its current formulation — is insufficient to prove every mathematical truth, let alone every truth of this complex world.

The age of computers

Rather than focus on what logic can't do, however, mathematicians and scientists have found endless ways to use logic. Foremost among these uses is the computer, which some experts (especially computer scientists) deem the greatest invention of the 20th century.

Hardware, the physical design of computer circuitry, uses *logic gates,* which mimic the basic functions of sentential logic — taking input in the form of electric current from one or two sources and outputting current only under given conditions.

For example, a NOT gate outputs current only when no current is flowing from its one input. An AND gate outputs only when current is flowing in from both of its inputs. And finally, an OR gate outputs only when current is flowing in from at least one of its two inputs.

Software, the programs that direct the actions of the hardware, is all written in *computer languages,* such as Java, C++, Visual Basic, Ruby, or Python. Although all computer languages have their differences, each contains a core of similarities, including a set of key words from sentential logic, such as *and, or, if . . . then,* and so on.

Check out Chapter 20 for more on how logic is used in computer hardware and software.

Searching for the final frontier

Will logic ever be sufficient to describe all of the subtleties of the human mind and the complexities of the universe? My guess is probably not — especially in its current formulation.

But, logic is an immensely powerful tool whose uses have only begun to be tapped into. And who knows? Even as I write this book, logicians are working to develop more expressive logical systems to expand the capabilities of mathematics and the sciences. Their efforts may yet produce inventions that surpass all current expectations and dreams.

Chapter 3

Just for the Sake of Argument

Simply put, logic is the study of how to tell a good argument from a bad one.

In daily life, most people use the word *argument* to describe anything from a tense disagreement to a shouting match. An argument, however, doesn't have to be heated or angry. The idea behind *logical argument* is simple: I want to convince you of something, so I point out some facts that you already agree with. Then, I show you how what I'm trying to prove follows naturally from these facts.

In this chapter, I explain what logic is and walk you through the elements of a logical argument. I also provide you with lots of examples, show you what logic *isn't*, and run through the various fields that use logic. When you're done, you may still find yourself on the wrong end of an argument, but at least you won't be accused of being illogical!

Defining Logic

Here's what you need to know about logic:

✔ Logic is the study of *argument validity*, which is whether a logical argument is *valid* (good) or *invalid* (bad).

✔ An *argument*, in logic, is a set of one or more *premises* followed by a *conclusion*, which are often connected by one or more *intermediate statements*.

> ✔ The premises and the conclusion are always *statements* — sentences that give information and that are either true or false.
>
> ✔ In a *valid argument*, if all of the premises are true, the conclusion must also be true.

When you put it all together, then, here's what you get:

> Logic is the study of how to decide the circumstances in which a set of true premises leads to a conclusion that is also true.

That's it! As you go through this book, keep this definition in mind. Write it down on an index card and paste it to the mirror in your bathroom. Every topic in logic relates to this central idea in some way.

Examining argument structure

A *logical argument* must have one or more premises followed by a conclusion. Here's an example of a logical argument:

> *Nick:* I love you.
>
> *Mabel:* Yes, I know.
>
> *Nick:* And you love me.
>
> *Mabel:* True.
>
> *Nick:* And people who love each other should get married.
>
> *Mabel:* OK.
>
> *Nick:* So we should get married.

This may not be the most romantic marriage proposal you've ever heard, but you get the idea. If Mabel really does agree with all three of Nick's statements, his argument should convince her to marry him.

Take a closer look at the structure of Nick's argument and you can see that it contains three premises and one conclusion. Boiling it down to its most basic form, Nick is saying:

Premises:

> I love you.
>
> You love me.
>
> People who love each other should get married.

Conclusion:

    We should get married.

The premises and the conclusion of an argument all have one thing in common: They are statements. A *statement* is simply a sentence that gives information.

For example, the following are all statements, though none yet fit into the premise or conclusion category:

- ✔ The capital of Mississippi is Jackson.
- ✔ Two plus two equals five.
- ✔ Your red dress is prettier than your blue one.
- ✔ Men are just like dogs.

In contrast, the following are *not* statements:

- ✔ A big blue Cadillac (not a complete sentence)
- ✔ Do you come here often? (a question)
- ✔ Clean your room right now. (a command)
- ✔ Golly! (an exclamation)

In logic, the information that a statement provides, and therefore the statement itself, may be either true or false. (This rule applies whether the statement you're talking about is being used as one of an argument's premises or as its conclusion.) This is called the *truth value* of that statement.

Because logicians deal with truth values all the time, they save on the ever-rising cost of ink by referring to the truth value as just the *value* of the statement. In this book, I use both terms, but they mean the same thing.

Sometimes you can easily verify the truth value of a statement. The value of a statement that says "The capital of Mississippi is Jackson" is *true* because Mississippi's capital is, in fact, Jackson. And two plus two is four, not five, so the value of the statement "Two plus two equals five" is *false.*

In other cases, the truth value of a statement is more difficult to verify. For example, how do you decide whether one dress really is prettier than another, or whether men really are just like dogs?

For the moment, don't worry about *how* you figure out whether a statement is true or false, or even *if* you can figure it out. I touch on this in the "The sound of soundness" section later in the chapter.

Looking for validation

In a good argument — or a *valid argument*, as logicians say — when all the premises are true, the conclusion must also be true.

Valid arguments are at the very heart of logic. Remember, in a logical argument, I want to convince you of something, so I point out facts that you already agree with (*premises*), then show you that what I'm trying to prove (the *conclusion*) follows from these facts. If the argument is valid, it's airtight, so the conclusion inevitably follows from the premises.

For example, suppose your professor says to you, "Everyone who studied did well on my midterm. And you studied, so you did well." Break this statement down into premises and a conclusion to see what you get:

Premises:

> If a student studied, then he or she did well on the midterm.
>
> You studied.

Conclusion:

> You did well on the midterm.

The preceding argument is an example of a valid argument. You can see that if both premises are true, then the conclusion must be true as well.

Now you can see why I also state in the preceding section that validity rests upon the structure of the argument. When this structure is missing, an argument is invalid even if all of its statements are true. For example, here is an invalid argument:

Premises:

> George Washington was the first U.S. president.
>
> Albert Einstein proposed the Theory of Relativity.

Conclusion:

> Bill Gates is the richest man in the world.

All of these statements happen to be true. But that doesn't mean the argument is valid. In this case, the argument is invalid because no structure is in place to ensure that the conclusion must follow from the premises. If Microsoft stock suddenly crashed and Bill Gates was penniless tomorrow, the premises would be true and the conclusion false.

Studying Examples of Arguments

Because arguments are so important to logic, I include in this section a few more examples so you can get a sense of how they work.

Some of these examples have arguments in which the first premise is in the form "*If . . . then . . . .*" *If* something happens, *then* something else will happen. You can think of this type of sentence as a slippery slide: Just step onto the *if* at the top of the slide and you end up sliding down into the *then* at the bottom.

Aristotle was the first person to study the form of arguments. For example:

Premises:

> All men are mortal.

> Socrates is a man.

Conclusion:

> Socrates is mortal.

He called this form of argument a *syllogism*. (See Chapter 2 for a closer look at early forms of logic by Aristotle and other thinkers.)

After you get the hang of how logical arguments work, the variations are endless. In Parts II through V of this book, you discover even more precise and useful ways to create, understand, and prove logical arguments. For now, though, these examples give you a taste of what valid arguments are all about.

Ice cream Sunday

Suppose that on Sunday, your son, Andrew, reminds you: "You said that if we go to the park on Sunday, we can get ice cream. And now we're going to the park, so that means we can get ice cream." His logic is impeccable. To show why this is so, here is Andrew's argument separated into premises and a conclusion:

Premises:

> If we go to the park, then we can get ice cream.

> We're going to the park.

Conclusion:

> We can get ice cream.

The first premise sets up the if-then slope, and the second is where you step onto it. As a result, you land inevitably in the conclusion.

Fifi's lament

Suppose one afternoon, you arrive home from school to hear your mother make the following argument: "If you cared about your dog, Fifi, you would take her for a walk every day after school. But you don't do that, so you don't care about her." Here's what you get when you break this argument down into its premises and a conclusion:

Premises:

> If you cared about Fifi, then you would walk her every day.
>
> You don't walk Fifi every day.

Conclusion:

> You don't care about Fifi.

The first premise here sets up an if-then slope. The second premise, however, tells you that you *don't* end up at the bottom of the slope. The only way this could happen is if you *didn't* step onto the slope in the first place. So, your mother's conclusion is valid — poor Fifi!

Escape from New York

Suppose that your friend Amy, describing where she lives, makes the following argument: "Manhattan is in New York. And Hell's Kitchen is in Manhattan. My apartment is in Hell's Kitchen, and I live there, so I live in New York." This argument also relies on the if-then slope, but doesn't contain the words "if" and "then." They're implied in the argument but not stated specifically. I make this clear below:

Premises:

> If something is in my apartment, then it's in Hell's Kitchen.
>
> If something is in Hell's Kitchen, then it's in Manhattan.
>
> If something is in Manhattan, then it's in New York.
>
> I live in my apartment.

Conclusion:

> I live in New York.

The if-then slope makes a conclusion apparent in this case. In this example, though, one slope leads to another, which leads to another. After you know that Amy lives in her apartment, you have no choice but to slide down the next three slopes to conclude that she lives in New York.

The case of the disgruntled employee

Suppose your wife, Madge, arrives home from work in a huff and says: "You can find three kinds of bosses in this world: the ones who pay you on time, the ones who apologize when they pay you late, and the ones who just don't value you as an employee. Well, my paycheck is late and my boss hasn't apologized, so I know that he doesn't value me."

Here's her argument:

Premises:

> A boss pays his employees on time or apologizes when he pays late or doesn't value you.
>
> My boss didn't pay me on time.
>
> My boss didn't apologize for the delay.

Conclusion:

> My boss doesn't value me.

This argument relies not on an if-then slope, but on a set of alternatives set up using the word "or." The first premise sets up the choice, while the second and third each eliminate an alternative. The conclusion is the only alternative that remains.

What Logic Isn't

Because it has been around for about 2,000 years, logic has had a chance to weave itself into the fabric of much of our culture. And *Star Trek's* Mr. Spock is only the tip of the iceberg.

Consider a few cultural stereotypes that involve logic: If you meet someone who's quiet and thoughtful, you may think that she is a *logical* person. When someone makes a rash or hasty decision — or one that you just don't agree with — you may accuse him of being *illogical* and advise him to *think logically* about what he's doing. On the other hand, if you find someone cold or detached, you may decide that he's *ruled by logic.*

Because logic is so loosely defined in common speech, people have all sorts of misconceptions about it. In some circles, logic is revered as the greatest human endeavor. In others, it's despised and looked at as an exercise for the ivory tower, which is remote from the challenges that face ordinary people in their daily lives.

Logic may have both a good and a bad reputation because people think it's something that it isn't. In this section, I pull out all the stops and show you exactly what logic is — and what it isn't.

You can use Table 3-1 as handy reference to compare what logic can and can't do.

Table 3-1	The Cans and Cannots of Logic
Logic Can't	**Logic Can**
Create a valid argument.	Critique a given argument for validity.
Tell you what is true or false in reality.	Tell you how to work with true and false statements.
Tell you if an argument is sound.	Tell you if an argument is valid.
Justify conclusions arrived at by induction.	Justify conclusions arrived at by deduction.
Make an argument rhetorically stronger (more convincing).	Provide the basis for rhetorical improvement.

Thinking versus logic

Reasoning, or even just plain thinking, is a hugely complex process that is only barely understood. And — to part company with Mr. Spock and his clan — logic is one piece of that reasoning process, but it doesn't make up the whole thing.

Think about the type of reasoning you might use to resolve a dispute between two children. You'd need to observe what is happening, relate this experience to past experiences, and try to figure out what might happen in the future. You might choose to act harshly by threatening to punish them. You might act kindly and reassure them. You might remain impartial and listen to both sides of the story. Or you might take a little from all of these approaches.

In short, you would have a lot of options and would somehow find a way to keep the peace (or maybe none of them would work, but that's a subject for *Parenting For Dummies*, not *Logic For Dummies*). All of these options involve thinking. Even sending both kids to their rooms without any attempt to understand what's going on requires the kind of thinking that humans do best (but that even the smartest dog cannot do).

However, if you limited yourself to logic when keeping the peace with your kids, you wouldn't get anywhere. Logic doesn't tell you what to do or how to act — it's not that kind of a tool. So when you respond to a situation by observing, thinking, and not letting emotion sway your actions, realize that you're calling on a whole host of abilities that are far more sophisticated (and also far less certain!) than logic.

Don't get logic and thinking confused. Logic is just one aspect of thinking, but it's an aspect that has been studied so rigorously that it's understood very well. In this way, logic is a lot more — and also a lot less — than just clear thinking unclouded by emotion.

Reality — what a concept!

Because with logic you're constantly working with true and false statements, you may think that the point of it is to tell you what is true and what is false — in other words, you may think that logic is supposed to tell you the nature of reality in objective terms. But, in fact, logic can't tell you what's objectively true or false — it can only tell you what's true or false relative to other statements that you already know (or believe) to be true or false.

For example, look at these two statements:

> New Jersey is a U.S. state.

> Stealing is always wrong.

The first statement seems simple to verify as true, but the second seems diffi-cult to verify at all. But, in either case, logic doesn't tell you right from wrong or even whether New Jersey is a state — you have to find other sources for those things. What logic *does* tell you is whether an argument is valid — that is, whether a set of true premises produces a true conclusion.

The only exception to this rule involves statements that are true in only a very limited and trivial sense. For example:

> All blue things are blue.

> Either Toonsis is a dog or he isn't a dog.

> If you're a salesman living in Denver, then you're a salesman.

Statements like these are called *tautologies*. Tautologies have an important role to play in logic, and you find out about that role in Chapter 6. For now, though, just understand that the main reason they're so reliably true is because they don't really convey any information about the world. As far as figuring out what in this crazy, complex world is true or false — you're on your own.

The sound of soundness

A *sound argument* is simply a valid argument plus the fact that the premises are true, making the conclusion true as well.

Soundness goes hand in hand with reality. But, because logic doesn't tell whether a statement is true or false, it also doesn't tell whether an argument is sound. Even a valid argument will produce a lousy conclusion if you give it a bunch of rotten premises. For example:

Premises:

Jennifer Lopez is on earth.

When the Messiah is on earth, the end times have arrived.

Jennifer Lopez is the Messiah.

Conclusion:

The end times have arrived.

Here you have a logically valid argument. However, is it sound or unsound? I personally have a hard time accepting the third premise as true, so I would have to say it's an unsound argument. And you may well agree. The point is this: Because logic won't take a side about whether that statement is true in reality, logic also won't help you decide whether the argument is sound.

Make sure you see the difference between a valid argument and a sound argument. A valid argument contains a big if: *If* all of the premises are true, then the conclusion must also be true. A sound argument is a valid argument with one further condition: The premises *really are true*, so, of course, the conclusion really is true as well.

Note that when you start off with an invalid argument, logic has a lot to say: If an argument is invalid it's always unsound as well. Figure 3-1 shows a little tree structure that may help you understand arguments better.

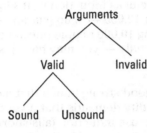

Figure 3-1:
A look at a
logic tree
can help
you tell
where your
arguments
fall.

Deduction and induction

Because *deduction* rhymes with *reduction*, you can easily remember that in deduction, you start with a set of possibilities and *reduce* it until a smaller subset remains.

For example, a murder mystery is an exercise in deduction. Typically, the detective begins with a set of possible suspects — for example, the butler, the maid, the business partner, and the widow. By the end of the story, he or she has reduced this set to only one person — for example, "The victim died in the bathtub but was moved to the bed. But, neither woman could have lifted the body, nor could the butler with his war wound. Therefore, the business partner must have committed the crime."

Induction begins with the same two letters as the word *increase*, which can help you remember that in induction, you start with a limited number of observations and *increase* that number by generalizing.

For example, suppose you spend the weekend in a small town and the first five people you meet are friendly, so you inductively conclude the following: "Everybody here is so nice." In other words, you started with a small set of examples and you increased it to include a larger set.

Logic allows you to reason deductively with confidence. In fact, it's tailor-made for sifting through a body of factual statements (*premises*), ruling out plausible but inaccurate statements (*invalid conclusions*), and getting to the truth (*valid conclusions*). For this reason, logic and deduction are intimately connected.

Deduction works especially well in math, where the objects of study are clearly defined and where little or no gray area exists. For example, each of the counting numbers is either even or odd. So, if you want to prove that a number is odd, you can do so by ruling out that the number is divisible by 2. See the "Whose Logic Is It, Anyway?" section for more examples of how and where logic is applied.

On the other hand, as apparently useful as induction is, it's logically flawed. Meeting 5 friendly people — or 10 or 10,000 — is no guarantee that the next one you meet won't be nasty. Meeting 10,000 people doesn't even guarantee that most people in the town are friendly — you may have just met all the nice ones.

Logic, however, is more than just a good strong hunch that a conclusion is correct. The definition of logical validity demands that if your premises are true, the conclusion is also true. Because induction falls short of this standard, it's considered the great white elephant of both science and philosophy: It looks like it may work, but in the end it just takes up a lot of space in the living room.

Rhetorical questions

Rhetoric is the study of what makes an argument cogent and convincing.

Take a look at the following argument:

Premises:

> Science can't explain everything in nature.

> Anything in nature can be explained by either science or by the existence of God.

Conclusion:

> God exists.

Even though this may or may not be a sound argument, it's a valid argument. However, it may not be a *cogent* argument.

What makes an argument cogent is close to what makes it sound (see the "The sound of soundness" section earlier in the chapter). In a sound argument, the premises are true beyond all doubt. In a cogent argument, the premises are true beyond a reasonable doubt.

The phrase "reasonable doubt" may remind you of a courtroom. This connection makes sense because lawyers make logical arguments that are intended to convince a judge or a jury of a certain conclusion. In order to be cogent, the lawyer's arguments must be valid and the premises of these arguments must be believable. (See the section "Tell it to the judge (law)" where I talk about the relevance of logic in a legal context.)

The science of induction

Induction is reasoning from a limited number of observations toward a general conclusion. A classic example: After observing that 2 or 10 or 1,000 ravens are black, you may decide that all ravens are black.

Inductive arguments, while they may be convincing, are not logically valid. (You never know, a white or pink raven may be out there somewhere.) But, although induction is logically unreliable, scientists seem to use it all the time with great success to explain all sorts of events in the universe. The philosopher Karl Popper took on the apparent contradiction, showing that induction isn't necessary for scientific discovery.

Popper's answer, in a nutshell, is that science doesn't really use induction to prove anything — it just looks that way. Instead, Popper says that scientists develop theories that fit their observations and then they *disprove* alternative theories. So, the last theory left standing becomes the accepted explanation until a better theory is found. (This may sound cavalier, as if any halfwit could come up with a crackpot theory and have it accepted. But, scientists are extremely adept at disproving plausible theories, so don't quit your day job.)

Even though some thinkers still question Popper's explanation, many find it a good explanation that resolves a centuries-old philosophical problem.

Considering that logic is concerned with validity rather than soundness, how does logic enter into the process of making an argument cogent? It does so by drawing a clear distinction between the underlying form of an argument and its content.

The form of an argument is what logic is all about. If the form doesn't function correctly, your argument falls apart. However, when it does function correctly, you can move on with confidence to the content of your argument.

Go back and take a look at the argument at the beginning of this section. I said that it's a valid argument but that it may not be cogent. Why? For one thing, the second premise is full of holes. Consider the following:

- ✔ What if science can't currently explain everything in nature but someday is able to?
- ✔ What happens if something other than science or the existence of God can explain some things?
- ✔ What if there is simply no explanation for everything we see in nature?

Each of these questions is a plausible exception to the argument. You need to address these exceptions if you want your argument to convince intelligent, clear-thinking people that your thinking is correct. In other words, you need to address the rhetorical issues within your argument.

In reality, however, appealing to the intelligence of an audience may be less effective than an appeal to powerful emotions and irrational beliefs. In such cases, a simple, flawed argument may be more convincing than a sound argument that the audience has trouble understanding. A shaky argument from a speaker whom the audience likes can be convincing where a brilliant but obnoxious display of intellect may be only alienating.

The study of what makes an argument cogent or convincing is useful and provocative, but it's also outside the scope of this book. From a purely logical standpoint, once an argument is valid, no further improvement is possible.

Whose Logic Is It, Anyway?

With all the restrictions placed upon it, you may think that logic is too narrow to be of much use. But this narrowness is logic's great strength. Logic is like a laser — a tool whose best use is not illumination, but rather focus. A laser may not provide light for your home, but, like logic, its great power resides in its precision. The following sections describe just a few areas in which logic is commonly used.

Pick a number (math)

Mathematics is tailor-made to use logic in all its power. In fact, logic is one of the three theoretical legs that math stands on. (The other two are set theory and number theory, if you're wondering.)

Logic and math work so well together because they're both independent from reality and because they're tools that are used to help people make sense of the world. For example, reality may contain three apples or four bananas, but the ideas of *three* and *four* are abstractions, even though they're abstractions that most people take for granted.

Math is made completely of such abstractions. When these abstractions get complicated — at the level of algebra, calculus, and beyond — logic can be called on to help bring order to their complexities. Mathematical ideas such as number, sum, fraction, and so on are clearly defined without exceptions. That's why statements about these ideas are much easier to verify than a statement about reality, such as "People are generally good at heart" or even "All ravens are black."

Fly me to the moon (science)

Science uses logic to a great advantage. Like math, science uses abstractions to make sense of reality and then applies logic to these abstractions

The sciences attempt to understand reality by:

1. Reducing reality to a set of abstractions, called a *model*.
2. Working within this model to reach a conclusion.
3. Applying this conclusion back to reality again.

Logic is instrumental during the second step, and the conclusions that science attains are, not surprisingly, logical conclusions. This process is most successful when a good correlation exists between the model and reality and when the model lends itself well to the type of calculations that logic handles comfortably.

The areas of science that rely most heavily on logic and math are the *quantifiable sciences,* such as physics, engineering, and chemistry. The *qualitative sciences* — biology, physiology, and medicine — use logic but with a bit less certainty. Finally, the *social sciences* — such as psychology, sociology, and economics — are the sciences whose models bear the least direct correlation to reality, which means they tend to rely less on pure logic.

Switch on or off (computer science)

Medicine used to be called the youngest science, but now that title has been handed over to computer science. A huge part of the success of the computer revolution rests firmly on logic.

Every action your computer completes happens because of a complex structure of logical instructions. At the hardware level — the physical structure of the machine — logic is instrumental in the design of complex circuits that make the computer possible. And, at the software level — the programs that make computers useful — computer languages based on logic provide for the endless versatility that sets the computer apart from all other machines.

See Chapter 20 for an in-depth discussion of logic as it relates to computers.

Tell it to the judge (law)

As with mathematics, laws exist primarily as sets of definitions: *contracts*, *torts*, *felonies*, *intent to cause bodily harm*, and so on. These concepts all come into being on paper and then are applied to specific cases and interpreted in the courts. A legal definition provides the basis for a legal argument, which is similar to a logical argument.

For example, to demonstrate copyright infringement, a plaintiff may need to show that the defendant published a certain quantity of material under his own name, for monetary or other compensation, when this writing was protected by a preexisting copyright.

These criteria are similar to the premises in a logical argument: If the premises are found to be true, the conclusion — that the defendant has committed copyright infringement — must also be true.

Find the meaning of life (philosophy)

Logic had its birth in philosophy and is often still taught as an offshoot of philosophy rather than math. Aristotle invented logic as a method for comprehending the underlying structure of reason, which he saw as the motor that propelled human attempts to understand the universe in the widest possible terms.

As with science, philosophy relies on models of reality to help provide explanations for what we see. Because the models are rarely mathematical, however, philosophy tends to lean more toward rhetorical logic than mathematical logic.

Part II
Formal Sentential Logic (SL)

The 5th Wave   By Rich Tennant

They're moving on to Chapter 2. That should daze and confuse them enough for us to finish changing the tire and get the heck out of here.

Logic Text Book Publishers

In this part . . .

*I*f you've ever glanced at a logic textbook (maybe the one that's been sitting on your desk accumulating dust since the semester began!), you may have wondered what those funky symbols → and ↔ mean. And, this is where you can find out. This part is exclusively about sentential logic, or SL for short, where all those symbols come into play.

In Chapter 4, you find out how to translate statements from English into SL using constants, variables, and the five logical operators. Chapter 5 gives you tips for when you enter the tricky world of evaluating SL statements to decide whether that statement is true or false. In Chapter 6, I introduce truth tables, which are powerful tools to find out all sorts of things about SL statements. Chapter 7 shows you how quick tables can replace truth tables as a fast way to solve problems. Finally, in Chapter 8, I introduce truth trees, which have all the advantages of truth tables and quick tables but none of the disadvantages.

Chapter 4

Formal Affairs

● ●

In This Chapter

▶ Introducing formal logic

▶ Defining the five logical operators

▶ Translating statements

● ●

*I*f you look at a few logical arguments, which you can do in Chapter 3, you may get the sneaking suspicion that they all have a lot in common. And you'd be right. Over the centuries, logicians have examined their fair share of argument examples, and what they found is that certain patterns of arguments come up again and again. These patterns can be captured with a small number of symbols, which can then be studied for their common features.

In this chapter, I introduce *formal logic,* a foolproof set of methods for determining whether an argument is valid or invalid. I show you how to represent statements with placeholders called *constants* and *variables,* and I introduce the five *logical operators* used for connecting simple statements into more complex ones.

The logical operators work very much like the familiar operators in arithmetic (like addition, subtraction, and so on), and I point out their similarities so that you can get comfortable with the new symbols. Finally, I show you how to translate statements in English into logical statements and vice versa.

Observing the Formalities of Sentential Logic

As I discuss in Chapter 2, *sentential logic* (or SL; also known as *propositional logic*) is one of two forms of classical formal logic. (The other form is *quantifier logic,* or QL, also known as *predicate logic;* I introduce SL in this chapter and dig deeper into it throughout Parts II and III. I save QL until Part IV.)

Logical arguments are communicated in language, but natural languages such as English tend to be imprecise. Words often have more than one meaning and sentences can be misinterpreted.

To help solve this problem, mathematicians and philosophers developed sentential logic, a language specifically designed to express logical arguments with precision and clarity. Because SL is a symbolic language, it has the added advantage of allowing for calculation according to precisely defined rules and formulas. Just as with mathematics, as long as you follow the rules correctly, the correct answer is guaranteed.

In the following sections, I introduce a few types of symbols that SL uses to achieve these goals.

Statement constants

If you ever spent a day in an algebra or pre-algebra class, you were probably exposed to that elusive little fellow known as x. Your teacher probably told you that x stood for a secret number and that it was your job to make x talk. He or she then showed you all sorts of sadistic ways of torturing poor little x until at last it broke down and revealed its true numerical identity. Oh, what fun that was.

Making letters stand for numbers is one thing mathematicians are really good at doing. So it isn't surprising that formal logic, which was developed by mathematicians, also uses letters as stand-ins. In the chapter introduction, I let on that logic uses *statements* rather than numbers, so it's logical to guess that in formal logic, letters stand for statements. For example:

Let K = Katy is feeding her fish.

Let F = The fish are joyfully flapping their little fins.

When a letter stands for a statement in English, the letter's called a *statement constant*. By convention, capital letters are used for constants.

For some reason, when it comes to constants, logicians tend to like the letters P and Q the most. Some folks say this is because P is the first letter in the word *proposition,* which means the same thing as statement, and Q just came along for the ride. My own personal theory is that after studying all that algebra in school, logicians were tired of using X and Y.

Statement variables

When logicians got the idea to make letters stand for statements, they just ran with it. They realized that they could use a letter for *any* statement, even a statement in SL. When letters are used in this way, they're called *statement variables*.

Throughout this book, I use variables to show overall patterns in SL and constants for nitty-gritty examples.

When a letter stands for a statement in SL, the letter's called a *statement variable*. By convention, small letters are used for variables. In this book, I use *x* and *y* almost exclusively, and occasionally use *w* and *z* when needed.

Truth value

As I cover in Chapter 3, every statement in logic has a *truth value* that's either true or false. In formal logic, *true* is shortened to **T** and *false* to **F**.

For example, consider the truth values of these two statement constants:

Let *N* = The Nile is the longest river in Africa.

Let *L* = Leonardo DiCaprio is the king of the world.

As it happens, it's true that the Nile is the longest river in Africa, so the truth value of *N* is **T**. And it so happens that Leonardo DiCaprio is *not* the king of the world, so the truth value of *L* is **F**.

Boolean algebra, the precursor to formal logic, uses the value *1* to represent **T** and *0* to represent **F**. These two values are still used in computer logic. (I discuss Boolean algebra in Chapter 14 and computer logic in Chapter 20.)

The Five SL Operators

SL has five basic operators, as you can see in Table 4-1. These *logical operators* are similar to the arithmetic operators in that they take values you give them and produce a new value. However, logical operators really only deal with two values: the truth values, **T** and **F**. In the sections that follow, I explain each of the operators in Table 4-1.

Table 4-1	The Five Logical Operators		
Operator	*Technical Name*	*What It Means*	*Example*
~	Negation	Not	~*x*
&	Conjunction	And	*x* & *y*
∨	Disjunction	Or	*x* ∨ *y*
→	Conditional	If . . . then	*x* → *y*
↔	Biconditional	If and only if	*x* ↔ *y*

Feeling negative

You can turn any statement into its opposite by adding or changing a few words. This is called *negating* the statement. Of course, negating a true statement turns it into a false statement, and negating a false statement makes it true. In general, then, every statement has the opposite truth value from its negation.

For example, I can tweak statement *N* below by simply inserting one little word:

N = The Nile is the longest river in Africa.

~*N* = The Nile is *not* the longest river in Africa.

The addition of *not* transforms the original statement into its opposite, which is the negation of that statement. Having established that the value of *N* is **T** (see the "Truth value" section earlier in the chapter), I can conclude that the value of its negation is **F**.

In SL, the negation operator is *tilde* (~). Consider another example of negation:

L = Leonardo DiCaprio is the king of the world.

~*L* = Leonardo DiCaprio is *not* the king of the world.

In this case, after establishing that the value of *L* is **F** (see the "Truth value" section earlier in the chapter), you also know that that value of ~*L* is **T**.

This information is simple to summarize in a table, as follows:

x	~x
T	F
F	T

Memorize the information in this table. You'll be using it a lot in the other chapters.

As you can see, in the table I use the variable *x* to stand for any SL statement. When the SL statement that *x* stands for is true, then ~*x* is false. On the other hand, when the statement that *x* stands for is false, then ~*x* is true.

Different books on logic may use the dash (–) or another symbol that looks like a sideways L for the not-operator rather than the tilde. I'm a tilde sort of guy, but whatever the symbol, it means the same thing.

Feeling even more negative

Although negation is only just the beginning, SL's little system of symbols has already grown more powerful than it looks. For example, given that the value of a new statement *R* is **T** and that its negation ~*R* is **F**, what can you say about ~~*R*?

If you guessed that the value of ~~*R* is **T**, give yourself a pat on the back. And as you do, notice that you were able to figure this out even though I didn't define *R*.

Witness the magic and power of logic. With just a few simple rules, you know that *any* statement of this form *must* be true, even if you don't know exactly what the statement is. This guarantee is similar to the idea that as soon as you know that 2 apples + 3 apples = 5 apples, you know that your result will be true no matter what you're counting: apples, dinosaurs, leprechauns — whatever.

Tabling the motion

Because *R* can take on only two possible truth values — **T** or **F** — you can organize the information about ~*R*, ~~*R*, and so forth in a table:

R	~R	~~R	~~~R
T	F	T	F
F	T	F	T

Tables of this kind are called *truth tables*. The first column contains the two possible truth values of *R*: **T** and **F**. The remaining columns give the corresponding values for the different related statements: ~*R*, ~~*R*, and so on.

Reading truth tables is pretty straightforward. Using the sample table above, if you know that the value of *R* is **F** and want to find out the value of ~~~*R*, you find the point where the bottom row crosses the last column. The truth value in this position tells you that when *R* is **F**, ~~~*R* is **T**.

Chapter 6 shows you how truth tables are a powerful tool in logic, but for now, I just use them to organize information clearly.

Displaying a show of ands

The symbol & is called the *conjunction operator* or, more simply, the *and-operator*. You can think of it as simply the word *and* placed between two statements, joining them together to make a new statement.

Check out this statement:

Albany is the capital of New York *and* Joe Montana was quarterback for the San Francisco 49ers.

Is this statement true or false? To decide, you need to recognize that this statement really contains two smaller statements: one about Albany and another about Joe Montana. Its truth value depends on the truth value of both parts.

Because both parts are true, the statement as a whole is true. Suppose, however, that one of the statements were false. Imagine some alternate universe where Albany is not the capital of New York or where Joe Montana was never the quarterback of the 49ers in the past. In either case, the truth value of the statement would be false.

In logic, you handle statements that involve the word *and* in a special way. First, each smaller statement is assigned a constant:

Let A = Albany is the capital of New York.

Let J = Joe Montana was the quarterback for the San Francisco 49ers.

Then, you connect the two constants as follows:

A & J

The truth value of this new statement is based on the truth values of the two parts that you've joined together. If *both* of these parts are true, then the whole statement is true. On the other hand, if *either* of these parts (or both of them) is false, then the whole statement is false.

For the &-operator, you can organize the truth values of x and y into a table:

x	y	x & y
T	T	T
T	F	F
F	T	F
F	F	F

Memorize the information in this table. Here's a quick way to remember it: An and-statement is true only when both parts of the statement are true. Otherwise, it's false.

Notice that the table for the &-operator above has four rows rather than just the two rows that were needed for the ~-operator (see the "Tabling the motion" section earlier in the chapter). The tables are different because the &-operator always operates on two variables, so its table has to track all four paired values for x and y.

Different books on logic may use a dot (·) or an inverted V for the and-operator rather than the ampersand. In other books, x & y is simply written xy. Whatever the convention, it means the same thing.

Digging for or

As with *and,* a statement may be made up of two smaller statements joined together with the word *or.* Logic provides an operator for the word *or:* The *disjunction operator,* or just plain *or-operator,* is ∨.

Take a look at this statement:

> Albany is the capital of New York *or* Joe Montana was quarterback for the San Francisco 49ers.

If you designate the first part of the statement as A and the second part as Q, you can connect the two constants A and Q as follows:

> $A \vee Q$

Is this statement true? Just as with a &-statement, when both parts of a ∨-statement are true, the entire statement is true. Therefore, the statement $A \vee Q$ has a truth value **T**. However, with a ∨-statement, even if only one part is true, the statement as a whole is still true. For example:

> Let A = Albany is the capital of New York.

> Let S = Joe Montana was the shortstop for the Boston Red Sox.

Now the statement $A \vee S$ means:

> Albany is the capital of New York *or* Joe Montana was shortstop for the Boston Red Sox.

Even though the second part of this statement is false, the whole statement is true because one part of it is true. Therefore, $A \vee S$ has a truth value of **T**.

But when *both* parts of a ∨-statement statement are false, the whole statement is false. For example:

> Let *Z* = Albany is the capital of New Zealand.

> Let *S* = Joe Montana was the shortstop for the Boston Red Sox.

Now the statement *Z* ∨ *S* means:

> Albany is the capital of New Zealand *or* Joe Montana was the shortstop for the Boston Red Sox.

This is a false statement because both parts of it are false. So *Z* ∨ *S* has a value of **F**.

For the ∨-operator, you can make a table with four rows that covers all possible combinations of truth values for *x* and *y*:

x	y	x ∨ y
T	T	T
T	F	T
F	T	T
F	F	F

Memorize the information in this table. A quick way to remember it is: An or-statement is false only when both parts of it are false. Otherwise, it's true.

In English, the word *or* has two distinct meanings:

- **Inclusive *or*:** When *or* means "this choice *or* that choice, *or both*"; the possibility that both parts of the statement are true is *included*. An example of an inclusive *or* is when a mom says "Before you go out, you need to clean your room *or* do your homework." Clearly, she means for her child to do one of these tasks *or both of them*.

- **Exclusive *or*:** When *or* means "this choice *or* that choice, *but not both*"; the possibility that both parts of the statement are true is *excluded*. An example of an exclusive *or* is when a mom says, "I'll give you money to go to the mall today *or* horseback riding tomorrow." She means that her child gets money for one of these treats, *but not both*.

English is ambiguous, but logic isn't. By convention, in logic, the ∨-operator is always *inclusive*. If both parts of a ∨-statement are true, the statement also is true.

Both the inclusive and exclusive *or* are used in the design of logic gates, an integral part of computer hardware. See Chapter 20 for more on computer logic.

Getting iffy

The symbol → is called the *conditional operator,* also known as the *if . . . then-operator* or just the *if-operator.* To understand how the →-operator works, take a look at this statement:

> If a wig is hanging off of a bedpost in the guest room, then Aunt Doris is visiting.

You can see that the statement contains two separate statements, each of which can be represented by a statement constant:

> Let W = A wig is hanging off of a bedpost in the guest room.
> Let D = Aunt Doris is visiting.

Then, connect the two with a new operator:

> $W \to D$

As with the other operators covered in this chapter, for the →-operator, you can make a table with four rows that covers all possible combinations of truth values for x and y:

x	y	x → y
T	T	T
T	F	F
F	T	T
F	F	T

Memorize the information in this table. A quick way to remember it is: An if-statement is false only when the first part of it is true and the second part is false. Otherwise, it's true.

Different books on logic may use a ⊂ for the if-operator rather than the arrow. Whatever the symbol, it means the same thing.

There's nothing random about the way the →-operator looks. The arrow points from left to right for an important reason: When an if-statement is true and the first part of it is true, the second part *must be true as well.*

To make this clear, I need a couple new constants:

Let *B* = You're in Boston.

Let *M* = You're in Massachusetts.

Now, consider the statement

$B \rightarrow M$

This statement means, "If you're in Boston, then you're in Massachusetts." Clearly, the statement is true, but why is this so? Because Boston is completely inside Massachusetts.

The converse of a statement

When you reverse an if-statement, you produce a new statement called the *converse* of that statement. For example, here's an if-statement followed by its converse:

If-statement: If you're in Boston, then you're in Massachusetts.

Converse: If you're in Massachusetts, then you're in Boston.

When an if-statement is true, it doesn't necessarily follow that its *converse* is true as well. Although the original statement above is true, the converse is false. You could be in Concord, Provincetown, or any number of other places in Massachusetts.

The inverse of a statement

When you negate both parts of an if-statement, you get another statement called the *inverse* of that statement. For example, compare the following:

If-statement: If you're in Boston, then you're in Massachusetts.

Inverse: If you're not in Boston, then you're not in Massachusetts.

When an if-statement is true, it doesn't necessarily follow that its *inverse* is true as well. Using the preceding example, even if you're not in Boston, you could still be somewhere else in Massachusetts.

The contrapositive of a statement

When you *both* reverse the order *and* negate the two parts of an if-statement, you get the *contrapositive* of the original statement. I know; I know. It's getting deep. But I have an example:

> **If-statement:** If you're in Boston, then you're in Massachusetts.

> **Contrapositive:** If you're not in Massachusetts, then you're not in Boston.

An if-statement and its contrapositive always have the same truth value. Using the example, given that the first part is true — you're not in Massachusetts — it's obvious that you can't be in Boston.

Although a statement and its contrapositive always have the same truth value, in practice proving the contrapositive of a statement is sometimes easier than proving the statement itself. (For more on proofs in SL, flip to Part III.) The converse of a statement always has the same truth value as the inverse of the same statement. This is because the converse and inverse are actually contrapositives *of each other*.

Getting even iffier

In SL, the if-and-only-if-operator (↔) is similar to the if-operator (see "Getting iffy" earlier in the chapter) but has more going on. The best way to understand the if-and-only-if-operator is first to establish an if-statement and then work your way along.

Consider this if-statement:

> *If* a wig is hanging off of a bedpost in the guest room, *then* Aunt Doris is visiting.

This statement says:

1. If you see a wig, then you *know* that Aunt Doris is here, *but*

2. If you see Aunt Doris, then you *can't be sure* that there is a wig.

This if-statement can be represented in SL as $W \rightarrow D$, with the arrow pointing in the direction of implication: Wig *implies* Doris.

Now consider this statement:

> A wig is hanging off of a bedpost in the guest room *if and only if* Aunt Doris is visiting.

This statement is similar to the previous one, but it extends things a bit further. In this case, the statement says:

1. If you see a wig, then you *know* that Aunt Doris is here, *and*

2. If you see Aunt Doris, then you *know* that there is a wig.

This statement can be represented in SL as $W \leftrightarrow D$, with the double arrow providing a clue to its meanings: *Both* wig implies Doris, *and* Doris implies wig.

Because the if-operator ($\rightarrow$) is also called the conditional operator, the if-and-only-if-operator ($\leftrightarrow$) is quite logically called the *biconditional operator*. Another shorter way of referring to it is the *iff-operator*. But this stuff can be tricky enough, so to be clear, I spell it out as if-and-only-if.

Don't confuse the if-and-only-if-operator ($\leftrightarrow$) with the if-operator ($\rightarrow$).

As with the other operators, for the $\leftrightarrow$-operator, you can make a table with four rows that covers all possible combinations of truth values for x and y:

x	y	$x \leftrightarrow y$
T	T	T
T	F	F
F	T	F
F	F	T

Memorize the information in this table. A quick way to remember it is: An if-and-only-if-statement is true only when both parts of it have the same truth value. Otherwise, it's false.

An important feature of the if-and-only-if-statement is that both parts of the statement are *logically equivalent,* which means that one can't be true without the other.

Check out two more examples of if-and-only-if-statements:

You're in Boston if and only if you're in Beantown.

A number is even if and only if you can divide it by two without a remainder.

The first statement is saying that Boston *is* Beantown. The second statement is pointing out the equivalence of its two parts — being an even number is equivalent to being evenly divisible by two.

Different books on logic may use ≡ for the if-and-only-if-operator rather than the double arrow. Whatever the symbol, it means the same thing.

How SL Is Like Simple Arithmetic

As I discuss in "The Five SL Operators" section earlier in this chapter, SL is a close cousin to math in that the operators for both disciplines take values you give them and produce a new value. But the similarity doesn't stop there. After you see a few other similarities, SL becomes much easier to understand.

The ins and outs of values

In arithmetic, each of the four basic operators turns two numbers into one number. For example:

$$6 + 2 = \mathbf{8} \qquad 6 - 2 = \mathbf{4} \qquad 6 \times 2 = \mathbf{12} \qquad 6 \div 2 = \mathbf{3}$$

The two numbers you start out with are called *input values*, and the number you end up with is the *output value*.

In each case, placing the operator between two input values (6 and 2) produces an output value (in boldface). Because there are two input values, these operators are called *binary operators*.

The minus sign also serves another purpose in math. When you place it in front of a positive number, the minus sign turns that number into a negative number. But when you place it in front of a negative number, the minus sign turns that number into a positive number. For example:

$$--4 = 4$$

In this case, the first minus sign operates on one input value (–4) and produces an output value (4). When used in this way, the minus sign is a *unary operator*, because there is only one input value.

In arithmetic, you have to worry about an infinite number of values. SL, however, only has two values: **T** and **F**. (For more on these truth values, jump back to the "Truth value" section earlier in this chapter.)

As with arithmetic, logic has four binary operators and one unary operator. In SL, the binary operators are &, ∨ →, and ↔, and the unary operator is ~. (Each of these operators is covered in the section "The Five SL Operators" earlier in this chapter.)

For both types of operators, the same basic rules apply in SL as in arithmetic:

> ✔ Stick a binary operator between any pair of input values and you get an output value.
>
> ✔ Place a unary operator in front of an input value and you get an output value.

For example, starting with a pair of input values, **F** and **T**, in that order, you can combine them by using the four binary operators as follows:

$$\textbf{F \& T = F} \qquad \textbf{F} \vee \textbf{T = T} \qquad \textbf{F} \rightarrow \textbf{T = T} \qquad \textbf{F} \leftrightarrow \textbf{T = T}$$

In each case, the operator produces an output value, which of course is either **T** or **F**. Similarly, placing the unary operator ~ in front of either input value **T** or **F** produces an output value:

$$\textbf{~F = T} \qquad \textbf{~T = F}$$

There's no substitute for substitution

Even if you've had only the slightest exposure to algebra, you know that letters can stand for numbers. For example, if I tell you that

$$a = 9 \text{ and } b = 3$$

you can figure out that

$$a + b = 12 \qquad a - b = 6 \qquad a \times b = 27 \qquad a \div b = 3$$

When you're working with constants in SL, the same rules apply. You just need to substitute the correct values (**T** or **F**) for each constant. For example, look at the following problem:

> Given that *P* is true, *Q* is false, and *R* is true, find the values of the following statements:

1. $P \lor Q$
2. $P \rightarrow R$
3. $Q \leftrightarrow R$

In problem 1, you substitute **T** for P and **F** for Q. This gives you **T** $\lor$ **F**, which equals **T**.

In problem 2, you substitute **T** for P and **T** for R. This gives you **T** $\rightarrow$ **T**, which equals **T**.

In problem 3, you substitute **F** for Q and **T** for R. This gives you **F** $\leftrightarrow$ **T**, which equals **F**.

Parenthetical guidance suggested

In arithmetic, parentheses are used to group numbers and operations together. For example:

$-((4 + 8) \div 3)$

In this expression, the parentheses tell you to solve 4 + 8 first, which gives you 12. Then, moving outward to the next set of parentheses, you solve 12 ÷ 3, giving you 4. Finally, the unary negation operator (–) changes this to –4.

In general, then, you start with the innermost pair of parentheses and work your way out. SL uses parentheses in the same way. For example, given that P is true, Q is false, and R is true, find the value of the following statement:

$\sim((P \lor Q) \rightarrow \sim R)$

Starting with the innermost pair of parentheses, $P \lor Q$ becomes **T** $\lor$ **F**, which simplifies to **T**. Then, moving outward to the next pair of parentheses, **T** $\rightarrow \sim R$ becomes **T** $\rightarrow$ **F**, which simplifies to **F**. And finally, the ~ appearing outside all the parentheses changes this **F** to **T**.

The process of reducing a statement with more than one value down to a single value is called *evaluating the statement*. It's an important tool that you find out more about in Chapter 5.

Lost in Translation

SL is a language, so after you know the rules, you can translate back and forth between SL and English . . . or Spanish or Chinese, for that matter — but let's stick to English! (However, if you're into Spanish or Chinese, check out *Spanish For Dummies* or *Chinese For Dummies*.)

The main strength of SL is that it's clear and unambiguous. These qualities make it easy to start with a statement in SL and translate it into English. That's why I call this translation direction the *easy way*. English, however, can be unclear and ambiguous. (See the section "Digging for *or*" earlier in the chapter for evidence that even the simple word *or* has several meanings, depending on how you use it.) Because you have to be very careful when translating sentences from English into SL, I call this the *not-so-easy way*.

Discovering how to translate statements in both directions will sharpen your understanding of SL by making the concepts behind all of these odd little symbols a lot clearer. And as you proceed into the next few chapters, if you start to get confused, just remember that every statement in SL, no matter how complex, could also be stated in English.

The easy way — translating from SL to English

Sometimes examples say it best. So here are examples of several ways to translate each type of operator. They're all pretty straightforward, so pick your favorite and run with it. Throughout this section, I use the following statement constants:

Let A = Aaron loves Alma.

Let B = The boat is in the bay.

Let C = Cathy is catching catfish.

Translating statements with ~

You can translate the statement $\sim A$ into English in any of the following ways:

> *It is not the case that* Aaron loves Alma.
>
> *It isn't true that* Aaron loves Alma.
>
> Aaron *does not love* Alma.
>
> Aaron *doesn't love* Alma.

Translating statements with &

Here are two ways to translate the SL statement $A \& B$:

> Aaron loves Alma *and* the boat is in the bay.
>
> *Both* Aaron loves Alma *and* the boat is in the bay.

Translating statements with ∨

Here are two ways to translate $A \vee C$:

Aaron loves Alma *or* Cathy is catching catfish.

Either Aaron loves Alma *or* Cathy is catching catfish.

Translating statements with →

You can translate the statement $B \rightarrow C$ in any of the following ways:

If the boat is in the bay, *then* Cathy is catching catfish.

The boat is in the bay *implies* Cathy is catching catfish.

The boat is in the bay *implies that* Cathy is catching catfish.

The boat is in the bay *only if* Cathy is catching catfish.

Translating statements with ↔

There's really only one way to translate the SL statement $C \leftrightarrow A$:

Cathy is catching catfish *if and only if* the boat is in the bay.

Translating more complex statements

For more complex statements, you can refer to the guidelines I discussed earlier in the chapter, in "How SL Is Like Simple Arithmetic." Simply translate statements step by step, starting from inside the parentheses. For example:

$(\sim A \mathbin{\&} B) \vee \sim C$

The part inside the parentheses is $(\sim A \mathbin{\&} B)$, which translates as

Aaron doesn't love Alma and the boat is in the bay.

Adding on the last part of the statement produces:

Aaron doesn't love Alma and the boat is in the bay *or* Cathy is not catching catfish.

Notice that although the sentence is technically correct, it's somewhat confusing because the parentheses are gone and everything runs together. A good way to clean it up is to translate it as

Either Aaron doesn't love Alma and the boat is in the bay *or* Cathy is not catching catfish.

The word *either* clarifies just how much the word *or* is meant to encompass. In contrast, the statement

$$\sim A \ \& \ (B \lor \sim C)$$

can be translated as

Aaron doesn't love Alma and *either* the boat is in the bay *or* Cathy is not catching catfish.

Now take a look at another example:

$$\sim(A \rightarrow (\sim B \ \& \ C))$$

Starting from the inner parentheses, you translate $(\sim B \ \& \ C)$ as

The boat is not in the bay and Cathy is catching catfish.

Moving to the outer parentheses, $(A \rightarrow (\sim B \ \& \ C))$ becomes this:

If Aaron loves Alma, then both the boat is not in the bay and Cathy is catching catfish.

Note the addition of the word *both* to make it clear that the original and-statement is in parentheses. Finally, add on the $\sim$ to get:

It isn't the case that if Aaron loves Alma, then both the boat is not in the bay and Cathy is catching catfish.

Okay, so it's a little hairy, but it makes about as much sense as a sentence like that ever will. You'll probably never have to translate statements much more complicated than these, but you should still reflect for a moment on the fact that SL can handle statements of any length with perfect clarity.

The not-so-easy way — translating from English to SL

Each of the four binary operators in SL ($\&$, $\lor$ $\rightarrow$, and $\leftrightarrow$) is a connector that joins a pair of statements together. In English, words that connect statements together are called *conjunctions*. Here are some examples of conjunctions:

although	if . . . then	or
and	either . . . or	so
but	neither . . . nor	therefore
however	nevertheless	though
if and only if	only if	

The ~-operator is usually translated into English as *not*, but it also can appear in other clever disguises, such as contractions (can't, don't, won't, and so on). With all this variety in language, an exhaustive account of how to turn English into SL would probably take ten books this size. What follows, then, is a slightly shorter account. In this section, I list some of the most common words and phrases, discuss each one, and then provide an example of how to translate it into SL.

First, though, I need to define some constants:

Let K = Chloe lives in Kentucky.

Let L = Chloe lives in Louisville.

Let M = I like Mona.

Let N = I like Nunu.

Let O = Chloe likes Olivia.

But, though, however, although, nevertheless . . .

Lots of words in English link statements together with the same logical meaning as the word *and*. Here are a few examples:

I like Mona, *but* I like Nunu.

Though I like Mona, I like Nunu.

Although I like Mona, I like Nunu.

I like Mona; *however*, I like Nunu.

I like Mona; *nevertheless*, I like Nunu.

Each of these words provides a slightly different meaning, but because you're just talking logic here, you translate them all as

M & N

After a statement has been translated from English to SL, the rules of SL take over. So here, just as with any &-statement, if either M or N is false, the statement M & N is false. Otherwise, it's true.

Neither . . . nor

The *neither . . . nor* structure negates both parts of the statement. For example:

I like *neither* Mona *nor* Nunu.

This statement means that *both* I don't like Mona *and* I don't like Nunu. Translate this statement into SL as

~M & ~N

Not . . . both

The *not . . . both* structure means that even though the statement as a whole is negated, either part by itself may not be negated. For example:

I do *not* like *both* Mona and Nunu.

This statement says that even though I don't like both women together, I may like one of them. So you translate this statement as

~(M & N)

. . . If . . .

You already know how to translate a statement that starts with the word *if*. But this becomes more confusing when you find the word *if* placed in the middle of a statement, as in:

I like Mona *if* Chloe likes Olivia.

To clarify the statement, just untangle it as follows:

If Chloe likes Olivia, I like Mona.

With the statement reorganized, you can see that the way to translate it is:

$O \rightarrow M$

. . . Only if . . .

This one is tricky until you think about it, and then it's really easy and you'll get it right every time. First, notice that the following statement is true:

Chloe lives in Louisville *only if* she lives in Kentucky.

This makes sense, because the only way that Chloe can live in Louisville is if she lives in Kentucky. Now, notice that the following statement is also true:

If Chloe lives in Louisville, then she lives in Kentucky.

This shows you that the two sentences are logically equivalent. So when faced with an . . . *only if* . . . between two parts of a statement, just realize that it's an if-statement that's already in the correct order. Translate it as

$L \rightarrow K$

. . . Or . . .

As I mention in the section "Digging for or" earlier in this chapter, this little word is big trouble. For example, it appears in this statement:

Chloe lives in Kentucky *or* Chloe likes Olivia.

Depending on how it's used, *or* can have two different meanings:

Chloe lives in Kentucky *or* Chloe likes Olivia, *or both.*

Chloe lives in Kentucky *or* Chloe likes Olivia, *but not both.*

Given *or*'s multiple personalities, my advice is this: When you see an *or* looking lonesome in a statement to be translated, it probably means that someone (like your professor) wants to make sure you know that, in logic, *or* always means *or . . . or both.* So translate the statement as

$K \vee O$

. . . Or . . . or both

This structure is clear and easy: It says what it means and means what it says. For example:

Chloe lives in Kentucky *or* Chloe likes Olivia, *or both.*

Translate this statement as

$K \vee O$

. . . Or . . . but not both

This structure has a clear meaning, but it's not so easy to translate. For example:

Chloe lives in Kentucky *or* Chloe likes Olivia, *but not both.*

In order to translate the words *but not both* into SL, you need to do some fancy logical footwork. As I discuss in the section "But, though, however, although, nevertheless . . ." earlier in the chapter, the word *but* becomes &, and the words *not both* are translated here as ~$(K \& O)$. Putting it all together, you translate the whole statement as

$(K \vee O) \,\&\, \sim(K \& O)$

Chapter 5

The Value of Evaluation

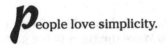eople love simplicity.

Have you ever read halfway through a movie review and then skipped to the end to find out whether the movie got a thumbs up or a thumbs down? Or, have you paged through a car magazine checking out how many stars every car received? I'm sure you never sat with a friend rating the guys or girls you both know on a scale of 1 to 10.

Movies, cars, guys, and girls are complicated. There's so much to understand. But people love simplicity. I'm sure you, like me, feel a sense of relief when you can reduce all of that complexity down to something small enough to carry in your pocket.

Logic was invented with this need in mind. In Chapter 4, you discover how to take a complicated statement in English and write it with just a few symbols in SL. In this chapter, you take this one step further by figuring out how to take a complicated statement in formal logic and reduce it to a single truth value: either **T** or **F**. Like the thumbs up or thumbs down of movie reviews, life doesn't get much simpler than that.

This conversion process is called *evaluating a statement* or *computing a statement's truth value*. Whichever term you use, it's one of the key skills you need in your study of logic. After you've mastered this process, a lot of locked doors will suddenly fling themselves open.

Value Is the Bottom Line

An important aspect of SL is that it allows you to simplify complex statements through the process of *evaluation*. When you evaluate an SL statement, you replace all of its constants with truth values (**T** or **F**) and then reduce the statement to a *single truth value*. When you think *evaluation*, remember that it means *to find the value of something*.

Correctly evaluating statements is probably the most important skill to master early on in your study of logic. Students who have trouble with this skill suffer for two reasons:

 ✔ It's time-consuming and frustrating if you don't know how to do it well.

 ✔ It's the first step you need to handle a bunch of other stuff, as you see later on in the book.

Here's the good news: Evaluating is a plug-and-chug skill. It doesn't require cleverness or ingenuity. You just need to know the rules of the game, and then you need to practice, practice, practice.

Those rules of the game for evaluating SL statements are a lot like the rules you already know for evaluating arithmetic statements. (Check out Chapter 4, where I outline other similarities between SL and arithmetic.) For example, take a look at this simple arithmetic problem:

$$5 + (2 \times (4 - 1)) = ?$$

To solve the problem, first evaluate what's inside the innermost set of parentheses. That is, because the value of $4 - 1$ is 3, you can replace $(4 - 1)$ with 3, and the problem becomes:

$$5 + (2 \times 3) = ?$$

Next, evaluate what's inside the remaining set of parentheses. This time, because the value of 2×3 is 6, you can make another replacement:

$$5 + 6 = ?$$

At this point, the problem is easy to solve. Because $5 + 6$ evaluates to 11, this is the answer. Through a series of *evaluations*, a string of numbers and symbols reduces to a single *value*.

Getting started with SL evaluation

Look at the following problem in SL:

Evaluate the statement $\sim(\sim P \rightarrow (\sim Q \,\&\, R)$

Here, the goal is the same as with an arithmetic problem: You want to evaluate the statement that you're given, which means you need to find its value.

In the arithmetic problem I discussed earlier in the chapter, though, you already knew the values of the four numbers (5, 2, 4, and 1). In the SL problem, you need to know the values of P, Q, and R. That is, you need to know an *interpretation* for the statement.

An *interpretation* of a statement is a fixed set of truth values for all of the constants in that statement.

For example, one possible interpretation for the statement is that P = **T**, Q = **F**, and R = **T**.

Remember that P, Q, and R are constants and that **T** and **F** are truth values. So when I write P = **T**, this is doesn't mean that these two things are equal. Instead, it is notation that means "the truth value of P is **T**.

You may not know for certain that this interpretation is correct, but you can still solve the problem under this interpretation — that is, under the assumption that it is correct. So, the full problem in this case would be:

Under the interpretation P = **T**, Q = **F**, R = **T**, evaluate the statement $\sim(\sim P \rightarrow (\sim Q \,\&\, R))$.

Now you can solve this problem. The first thing to do is replace each constant with its truth value:

$\sim(\sim\textbf{T} \rightarrow (\sim\textbf{F} \,\&\, \textbf{T}))$

After you replace the constants in an SL statement with their truth values, technically you don't have a statement any more, and the purists may scowl. But while you are learning about evaluation, turning SL statements into expressions of this kind is helpful.

The second and third ~-operators are directly linked to truth values, which makes them easy to evaluate because the value of ~**T** is **F** and the value of ~**F** is **T**. (Check out Chapter 4 for a refresher on working with the logical operators that you find in this chapter.) So, you can rewrite the expression:

$\sim(\textbf{F} \rightarrow (\textbf{T} \,\&\, \textbf{T}))$

Parentheses in SL work just as they do in arithmetic. They break up an expression so that it's clear what you need to figure out first. In this case, the innermost set of parentheses contains **T & T**. And because the value of **T & T** is **T**, this expression simplifies to:

$$\sim(F \rightarrow T)$$

Now, evaluate what's in the remaining set of parentheses. Because the value of **F → T** is **T**, the expression simplifies further to:

$$\sim(T)$$

At this point, it's easy to see that **~T** evaluates to **F**, which is the answer. The result here is similar to the result of the arithmetic problem: You started with a complex statement and evaluated it by finding its value, which in logic is always either **T** or **F**.

Stacking up another method

The evaluation method I used in the previous section works for all statements in SL, no matter how complex they are. In this next example, I use the same method with a small cosmetic change: Instead of rewriting the entire equation at every step, I just accumulate truth values as I go along. Here's a new problem:

Evaluate $\sim(\sim P \,\&\, (\sim Q \leftrightarrow R))$ using the interpretation $P = F$, $Q = T$, and $R = T$.

The first step is to replace the constants by their truth values. In the previous section's example, I rewrote the whole equation. This time, just place the truth value for each constant directly below it:

$$\sim(\sim P \,\&\, (\sim Q \leftrightarrow R))$$
$$\quad\;\; F \qquad\;\; T \quad\;\; T$$

This example has two ~-operators that immediately precede constants. These operators are very easy to work with: Simply put the correct value under each operator. As you can see below, the new values are larger, and the underlined values next to them show you where these values came from.

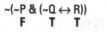

At this stage, don't try to evaluate any ~-operator that immediately precedes an open parenthesis. Because this operator negates *everything* inside the parentheses, you have to wait until you know the value of everything inside before you can put the operator into play.

Now you can work on what's inside the parentheses. Be sure to start on the inside set. The operator you're evaluating here is the ↔-operator. On one side of it, the value of ~Q is **F**. On the other side, the value of R is **T**. That gives you **F ↔ T**, which evaluates to **F**. Place this value, directly under the operator you just evaluated, which is the ↔-operator. Doing so allows you to see that the value of everything inside the parentheses is **F**:

$$~(~P \& (~Q \leftrightarrow R))$$
$$\text{TF} \quad \text{FT } \textbf{F} \text{ T}$$

Now move outward to the next set of parentheses. Here, you're evaluating the &-operator. On one side, the value of ~P is **T**. On the other side, the value of everything inside the inner parentheses (that is, the value under the ↔) is **F**. That gives you **T & F**, which evaluates to **F**. Place this value under the &-operator:

$$~(~P \& (~Q \leftrightarrow R))$$
$$\text{TF } \textbf{F} \text{ FT } \text{F} \text{ T}$$

The final step is to evaluate the entire statement. The operator you're evaluating now is the ~-operator. This negates everything inside the parentheses — meaning the value under the &-operator, which evaluates to **F**. Place this value under the ~-operator:

$$~(~P \& (~Q \leftrightarrow R))$$
$$\textbf{T} \text{ TF } \text{F} \text{ FT } \text{F} \text{ T}$$

Now, everything is evaluated, so this last value, **T**, is the value of the whole statement. In other words, when P = **F**, Q = **T**, and R = **T**, the statement ~(~P & (~Q ↔ R)) evaluates to **T**. As you can see, evaluation allows you to turn a great deal of information into a single truth value. It doesn't get much simpler than that!

Making a Statement

Now that you've had a taste of evaluation, I'm going to give you a closer look at how SL statements work. After you understand a little bit more about statements, you'll find that they practically evaluate themselves.

Identifying sub-statements

A *sub-statement* is any piece of a statement that can stand on its own as a complete statement.

For example, the statement $P \lor (Q \& R)$ contains the following two sub-statements that can stand on their own as complete statements:

- ✔ $Q \& R$
- ✔ P

And here is an example of a piece of the statement $P \lor (Q \& R)$ that is *not* a sub-statement: $\lor (Q \&$. Even though this is a piece of the statement, it obviously isn't a complete statement in its own right. Instead, it's just a meaningless string of SL symbols. (In Chapter 13, you discover the fine points of how to tell a statement from a string of symbols.)

When you evaluate a statement, you begin by placing values on the smallest possible sub-statements, which are the individual constants.

For example, suppose you want to evaluate the statement $P \lor (Q \& R)$ based on the interpretation $P = $ **T**, $Q = $ **T**, $R = $ **F**. You begin by placing the truth value below each constant:

$$P \lor (Q \& R)$$
$$\text{T} \quad \text{T} \quad \text{F}$$

Now you can evaluate the larger sub-statement $Q \& R$:

$$P \lor (Q \& R)$$
$$\text{T} \quad \text{T} \, \text{F} \, \text{F}$$

Finally, you can evaluate the whole statement:

$$P \lor (Q \& R)$$
$$\text{T} \, \text{T} \quad \text{T} \, \text{F} \, \text{F}$$

As you can see, evaluation of a long statement works best when you break it down piece by piece into sub-statements that can be evaluated more easily.

Scoping out a statement

Once you know what a sub-statement is, it's easy to understand the scope of an operator. The *scope* of an operator is the smallest sub-statement that includes the operator in question.

For example, take the statement $(P \rightarrow (Q \vee R)) \leftrightarrow S$. Suppose you want to know the scope of the $\vee$-operator. Two possible sub-statements that include this operator are $P \rightarrow (Q \vee R)$ and $Q \vee R$. The shorter of the two is $Q \vee R$, so this is the scope of the operator.

You can also think of an operator's scope as the *range of influence* that this operator holds over the statement.

To illustrate the range of influence, I've underlined the scope of the $\vee$-operator in the following statement:

$$(P \rightarrow (\underline{Q \vee R})) \leftrightarrow S$$

This shows that the $\vee$-operator affects, or influences, the constants Q and R, but *not* the constants P or S.

In contrast, I've underlined the scope of the $\rightarrow$-operator in the same statement:

$$(\underline{P \rightarrow (Q \vee R)}) \leftrightarrow S$$

This shows that the $\rightarrow$-operator's range of influence includes the constant P and the sub-statement $(Q \vee R)$, but not the constant S.

Before you can evaluate an operator, you need to know the truth value of every other constant and operator in its scope. And, once you understand how to find the scope of an operator, it's easy to see why you need to begin evaluating a statement from inside the parentheses.

For example, in the statement $(P \rightarrow (Q \vee R)) \leftrightarrow S$, the $\vee$-operator is within the scope of the $\rightarrow$-operator. This means that you can't evaluate the $\rightarrow$-operator until you know the value of the $\vee$-operator.

Be careful when figuring out the scope operators in statements with ~-operators. The scope of a ~-operator is always the smallest sub-statement that immediately follows it. When a ~-operator is in front of a constant, its scope includes only that constant. You can think of a ~-operator in front of a constant as being bound to that constant. For example, the scope of the first ~-operator is underlined:

$$\underline{\sim P} \,\&\, \sim(Q \,\&\, R)$$

In contrast, when a ~-operator is in front of an open parenthesis, its scope is everything inside that set of parentheses. For example, the scope of the second ~-operator is underlined:

~P & ~(Q & R)

Similarly, you might underline the scope of the ∨-operator in the statement ~(P ∨ Q) as follows:

~(P ∨ Q)   WRONG!

In this case, the ~-operator is outside the parentheses, so it's outside the scope of the ∨-operator.

~(P ∨ Q)   RIGHT!

So when you're evaluating this statement, *first* evaluate the sub-statement P ∨ Q, and *then* evaluate the entire statement.

The main attraction: Finding main operators

The *main operator* is the most important operator in a statement, for the following reasons:

- ✔ **Every SL statement has just one main operator.**

- ✔ **The scope of the main operator is the whole statement.** Thus, the main operator affects every other constant and operator in the statement.

- ✔ **The main operator is the last operator that you evaluate.** This fact makes sense when you think about it: Because the scope of the main operator is *everything else* in the statement, you need to evaluate everything else before you can evaluate the main operator.

  For example, suppose you want to evaluate (P → (Q ↔ R)) & S under a given interpretation. First you need to evaluate the sub-statement Q ↔ R to get the value of the ↔-operator. Doing this allows you to evaluate P → (Q ↔ R) to get the value of the →-operator. And finally, you can evaluate the whole statement, which gives you the value of the statement's main operator, the &-operator. (I show you how to find the main operator later in this section.)

- ✔ **The main operator's value is the same as the value of the statement itself.**

- ✔ **The main operator falls outside of all parentheses,** *except* **when the whole statement includes an extra (and removable) set of parentheses.** I explain this in detail in the remainder of this section.

Because the main operator is so important, you need to be able to pick it out of any statement. Doing so is usually quite simple with a few quick rules of thumb. Every SL statement falls into one of the three cases I outline in the following sections. If you come across one that doesn't, it isn't well-formed, which means it really isn't a statement at all. I discuss this more in Chapter 14. For now, though, any statement you run across will have a main operator that you can find without difficulty.

When only one operator is outside the parentheses

Sometimes, it's easy to find the main operator because it's the *only* operator outside all parentheses. For example, take a look at this statement:

$$(P \lor \sim Q) \,\&\, (R \to P)$$

The main operator here is the &-operator. Similarly, check out this statement:

$$\sim (P \,\&\, (Q \leftrightarrow R))$$

The main operator here is the ~-operator.

When no operator is outside the parentheses

If you find *no* operator outside the parentheses, you have to remove a set of parentheses. For example, in the following statement, the outer set of parentheses really isn't necessary:

$$((\sim P \leftrightarrow Q) \to R)$$

But, removing them gives you this:

$$(\sim P \leftrightarrow Q) \to R$$

Now, the only operator outside the parentheses is the →-operator, which is indeed the main operator.

In this book, I avoid using unnecessary parentheses because they take up space and don't add anything useful to a statement. In Chapter 14, I discuss the nitty-gritty details of why a statement may contain extra parentheses.

When more than one operator is outside the parentheses

In some statements, you may find more than one operator outside the parentheses. For example:

$$\sim(\sim P \to Q) \lor \sim(P \to Q)$$

When there is more than one operator outside the parentheses, the main operator is always the one that *isn't* a ~-operator.

In the preceding example, the main operator is the ∨-operator.

Eight Forms of SL Statements

In SL, a variable can stand for an entire statement (or sub-statement). You can use variables to classify SL statements into eight different *statement forms*, which are generalized versions of SL statements. Table 5-1 shows the eight basic statement forms.

Table 5-1	The Eight Forms of SL Statements
Positive Forms	*Negative Forms*
$x \& y$	$\sim(x \& y)$
$x \vee y$	$\sim(x \vee y)$
$x \rightarrow y$	$\sim(x \rightarrow y)$
$x \leftrightarrow y$	$\sim(x \leftrightarrow y)$

To see how these statements work, here are three statements whose main operators are all &-operators:

$P \underline{\&} Q$

$(P \vee \sim Q) \underline{\&} \sim(R \rightarrow S)$

$(((\sim P \leftrightarrow Q) \rightarrow R) \vee (\sim Q \& S)) \underline{\&} R$

Parts of the whole

Personally, I find all of the following terminology for the various parts of a statement to be a bit over the top. If your professor wants you to know these terms, you're going to have to memorize them. But for me, the important thing is that when you're given an SL statement, you can find the main operator and pick out which of the eight forms it's in. When it becomes important to speak about the various parts of a statement, it's just as easy to say "first part" and "second part."

Here are some quick rules of thumb:

✔ When a statement is in the form $x \& y$, it's a &-statement, which is also called a *conjunction*. In this case, both parts of the statement are called *conjuncts*.

✔ When a statement is in the form $x \vee y$, it's a $\vee$-statement, which is also called a *disjunction*. In this case, both parts of the statement are called *disjuncts*.

✔ When a statement is in the form $x \rightarrow y$, it's a $\rightarrow$-statement, which is also called an *implication*. In this case, the first part of the statement is called the *antecedent*, and the second part is called the *consequent*.

✔ When a statement is in the form $x \leftrightarrow y$, it's a $\leftrightarrow$-statement, which is also called a *double implication*.

Although all of these statements are obviously different, you can represent each of them using the following statement form:

$x \& y$

For example, in the statement $P \& Q$, the variable x stands for the sub-statement P, and the variable y stands for the sub-statement Q. Similarly, in the statement $(P \vee {\sim}Q) \underline{\&} {\sim}(R \to S)$, x stands for the sub-statement $(P \vee {\sim}Q)$, and y stands for the sub-statement ${\sim}(R \to S)$. And finally, in the statement $(((({\sim}P \leftrightarrow Q) \to R) \vee ({\sim}Q \& S)) \& R$, x stands for $((({\sim}P \leftrightarrow Q) \to R) \vee ({\sim}Q \& S))$ and y stands for R.

When a statement's main operator is one of the four binary operators ($\&$, $\vee$, $\to$, or $\leftrightarrow$), its statement form is one of the four positive forms in Table 5-1. However, when a statement's main operator is the ${\sim}$-operator, its form is one of the negative forms in Table 5-1. To find out which one, you need to look at the operator with the next-widest scope. For example:

${\sim}((P \to Q) \leftrightarrow (Q \vee R))$

In this case, the main operator is the ${\sim}$-operator. The $\&$-operator has the next-widest scope, covering everything inside the parentheses. So, you can represent this statement using this statement form:

${\sim}(x \leftrightarrow y)$

This time, the variable x stands for the sub-statement $(P \to Q)$, and the variable y stands for the sub-statement $(Q \vee R)$.

Learning to recognize the basic form of a given statement is a skill you'll use in later chapters. For now, be aware that every statement can be represented by just one of the eight basic statement forms.

Evaluation Revisited

After reading about the new concepts in this chapter, you'll probably find that evaluation makes more sense. You're less likely to make mistakes because you understand how all the pieces of the statement fit together.

For example, suppose you want to evaluate ${\sim}({\sim}(P \vee Q) \& ({\sim}R \leftrightarrow S))$ under the interpretation $P = \textbf{T}$, $Q = \textbf{F}$, $R = \textbf{F}$, and $S = \textbf{T}$. It looks hairy, but you should be up for the challenge!

Before you begin, look at the statement. It's in the form ${\sim}(x \& y)$, with the first part of the statement being ${\sim}(P \vee Q)$ and the second part being $({\sim}R \leftrightarrow S)$. You need to get the truth value of both of these parts before you can evaluate the

&-operator. Only then can you evaluate the first ~-operator, which is the statement's main operator.

You start out by placing the truth values under the appropriate constants:

$$\sim(\sim(P \lor Q) \& (\sim R \leftrightarrow S))$$
$$\quad\quad T \quad F \quad\quad F \quad T$$

Now, you can write in the value of the ~-operator in front of the constant R:

$$\sim(\sim(P \lor Q) \& (\sim R \leftrightarrow S))$$
$$\quad\quad T \quad F \quad\quad T_F \quad T$$

At this point, you can get the value of both the V-operator and the ↔-operator:

$$\sim(\sim(P \lor Q) \& (\sim R \leftrightarrow S))$$
$$\quad\quad I\,T\,F \quad\quad IF \; T \; I$$

You may be tempted at this point to evaluate the &-operator, but first you need the value of the sub-statement $\sim(x \lor y)$, which means that you need to get the value of the ~-operator:

$$\sim(\sim(P \lor Q) \& (\sim R \leftrightarrow S))$$
$$\quad F\,I\,F \quad\quad TF \; T \; T$$

Now you can evaluate the &-operator:

$$\sim(\sim(P \lor Q) \& (\sim R \leftrightarrow S))$$
$$\quad F\,T\,T\,F \; F \quad TF \; I \; T$$

And finally, after you've evaluated every other operator in the statement, you can evaluate the main operator:

$$\sim(\sim(P \lor Q) \& (\sim R \leftrightarrow S))$$
$$T\,F\,T\,I\,F \; F \quad TF \; I \; T$$

The truth value of the main operator is the value of the whole statement, so you know that under the given interpretation, the statement is true.

Chapter 6

Turning the Tables: Evaluating Statements with Truth Tables

. . . .

In This Chapter

▶ Creating and analyzing truth tables

▶ Knowing when statements are tautologies, contradictions, or contingent

▶ Understanding semantic equivalence, consistency, and validity

. . . .

*I*n this chapter, you discover one of the most important tools in sentential logic (SL): the *truth table*. Truth tables allow you to evaluate a statement under every possible interpretation, which in turn allows you to make general conclusions about a statement even when you don't know the truth values of its constants.

Truth tables open up vast new logical territory. First of all, truth tables are an easy way to find out whether an argument is valid or invalid — a central question in logic. But beyond this, truth tables make it possible to identify *tautologies* and *contradictions*: Statements in SL that are always true or always false.

You can also use truth tables to decide whether a set of statements is *consistent* — that is, whether it's possible that all of them are true. Finally, with truth tables, you can figure out whether two statements are *semantically equivalent* — that is, whether they have the same truth value in all possible cases.

This is where the rubber meets the road, so fasten your seat belts!

Putting It All on the Table: The Joy of Brute Force

Sometimes, solving a problem requires cleverness and ingenuity. Before you can get the answer, you need to have that "Aha!" moment that allows you to see things in a whole new way. "Aha!" moments can be exhilarating, but also frustrating, especially if you're working on an exam under time pressure and the "Aha!" is nowhere in sight.

Truth tables are the antidote to "Aha!" They rely on a method that mathematicians informally call *brute force*. In this type of approach, instead of trying to find that one golden path to success, you doggedly exhaust all possible paths. Brute force methods can be time-consuming, but at the end of the day, you always find the answer you're looking for.

Here's how it works. Suppose on your first exam you're given this cryptic question:

> What can you say about this statement: $P \rightarrow (\sim Q \rightarrow (P \,\&\, \sim Q))$? Justify your answer.

With the clock ticking, you can probably think of a lot of things you want to say about the statement, but they're not going to get you any points on the test. So, you stare at it, waiting for the "Aha!" to arrive. And suddenly it does. You finally think about it this way:

> This statement is just stating the obvious: "If I assume P is true and then assume that Q is false, I can conclude that P is true and Q is false." So, the statement is *always* true.

That wasn't so bad. Still, what if the "Aha!" never arrived? What if you weren't sure how to "justify your answer?" Or, worst of all, what if the question you faced looked like this:

> What can you say about this statement: $((\sim P \vee Q) \rightarrow ((R \,\&\, \sim S) \vee T)) \rightarrow (\sim U \vee ((\sim R \vee S) \rightarrow T)) \rightarrow ((P \,\&\, \sim U) \vee (S \rightarrow T))$? Justify your answer.

These are times for brute force, and truth tables are the ticket.

Baby's First Truth Table

A truth table is a way of organizing information in SL. It allows you to set up every possible combination of values for every constant and see what happens in each case. You've already seen some small truth tables in Chapter 4: As I introduced each of the five SL operators, I included a truth table showing how to interpret all possible input values to obtain an output value.

In the following section, I show you how to set up, fill in, and draw conclusions from a truth table by working through the statement $P \rightarrow (\sim Q \rightarrow (P \, \& \sim Q))$.

Setting up a truth table

A truth table is a way to organize *every possible interpretation* of a statement into horizontal rows, allowing you to evaluate the statement under all of these interpretations.

Setting up a truth table is a simple four-step process:

1. **Set up the top row of the table with each constant on the left and the statement on the right.**

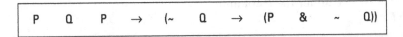

P	Q	P	→	(~	Q	→	(P	&	~	Q))

2. **Determine the number of additional rows that your table needs, based on the number of constants in the statement.**

A truth table needs one row for every possible interpretation for the given statement. To figure out how many rows you need, multiply the number two by itself one time for every constant in that statement. Because the statement $P \rightarrow (\sim Q \rightarrow (P \, \& \sim Q))$ has two constants, you'll need four rows.

To make sure that you're on the right track, check out Table 6-1, which contains the number of rows you need for statements with one to five constants.

Table 6-1	Number of Constants and Rows in Truth Table	
Number of Constants	Constants	Number of Interpretations (Rows in Truth Table)
1	P	2
2	P and Q	$2 \times 2 = 4$
3	P, Q, and R	$2 \times 2 \times 2 = 8$
4	P, Q, R, and S	$2 \times 2 \times 2 \times 2 = 16$
5	P, Q, R, S, and T	$2 \times 2 \times 2 \times 2 \times 2 = 32$

3. **Set up the constant columns so that with every possible combination of truth values is accounted for.**

A good way to fill in these columns with Ts and Fs is to start in the right-hand constant column (in this case, the Q column) and fill that column in by alternating Ts and Fs — **T F T F** — all the way down. Then move one column to the left and fill that column in, alternating the Ts and Fs by twos — **T T F F**. If there are more columns (for example, you have three or four constants in a statement), continue alternating by fours (**T T T T F F F F. . .**), eights (**T T T T T T T T F F F F F F F F. . .**), and so on.

So, in this example, alternate **T F T F** under the Q column, and then **T T F F** under the P column:

P	Q	P	→	(~	Q	→	(P	&	~	Q))
T	T									
T	F									
F	T									
F	F									

4. **Draw horizontal lines between rows, and draw vertical lines separating all constants and operators in the statement.**

P	Q	P	→	(~	Q	→	(P	&	~	Q))
T	T									
T	F									
F	T									
F	F									

I suggest this step for three reasons. First, the table starts out neat so you don't get confused. Second, you'll know the table is finished when all the little boxes are filled in with **Ts** and **Fs**. And third, the completed table is clear and readable. (If you draw the lines nicely with a ruler, you'll melt the heart of even the iciest professor.)

You don't need columns for parentheses, but be sure to bunch each parenthesis in with the constant or operator that immediately *follows* it, except for those at the end of the statement.

Filling in a truth table

Every row of your truth table accounts for a different interpretation of the statement. Filling in the truth table now becomes simply the process of evaluating the statement under every interpretation (in this case, under all four interpretations).

In Chapter 4, I discuss how to evaluate a statement from the inside out. The rules are still the same in this instance, but now you need to work every step of the process on every row of the table.

The step-by-step procedure that follows shows you how to work down the columns because it's easier and faster than trying to evaluate row by row.

As I work through the steps, note that each column I've just filled in is underlined, and each column I used during this step is in boldface.

1. **Copy the value of each constant into the proper statement columns for those constants.**

P	Q	P	→	(~	Q	→	(P	&	~	Q))
T	T	T			T		T			T
T	F	T			F		T			F
F	T	F			T		F			T
F	F	F			F		F			F

Just copying. Pretty simple, right?

2. **In each of the columns that has a ~-operator directly in front of a constant, write the corresponding negation of that constant in each row.**

P	Q	P	→	(~	Q	→	(P	&	~	Q))
T	T	T		F	T		T		F	T
T	F	T		T	F		T		T	F
F	T	F		F	T		F		F	T
F	F	F		T	F		F		T	F

WARNING! Make sure that every ~-operator is in front of a constant. If it's in front of an open parenthesis, it negates the entire value of everything inside the parentheses. In that case, you have to wait until you know this value.

As you can see, this step is not much more difficult than the previous one.

3. **Starting with the innermost set of parentheses, fill in the column directly below the operator for that part of the statement.**

Step 3 is really the meat and potatoes of truth tables. The good news here is that with practice you're going to get really fast at this stage of filling in the tables.

The innermost parentheses in this example contain the statement *P* & ~*Q*. The operator you're using to evaluate is the &-operator, and the two input values are in the columns under *P* and the ~-operator.

For example, in the first row, the value under *P* is **T** and the value under the ~-operator is **F**. And **T & F** evaluates to **F**, so write this value in this row directly under the &-operator.

Repeat this step for the other three rows and your table should look like this:

P	Q	P	→	(~	Q	→	(P	&	~	Q))
T	T	T		F	T		T	F	F	T
T	F	T		T	F		T	T	T	F
F	T	F		F	T		F	F	F	T
F	F	F		T	F		F	F	T	F

4. **Repeating Step 3, work your way outward from the first set of parentheses until you've evaluated the main operator for the statement.**

   Moving outward to the next set of parentheses, the operator you're using to evaluate is the →-operator inside the outermost parentheses. The two input values are in the columns under the first ~-operator and the &-operator.

   For example, in the first row, the value under the ~-operator is **F** and the value under the &-operator is **F**. And **F → F** evaluates to **T**, so write this value in the row directly under the →-operator.

   Complete the column and your table should look like this:

P	Q	P	→	(~	Q	→	(P	&	~	Q))
T	T	T		F	T	T	T	F	F	T
T	F	T		T	F	T	T	T	T	F
F	T	F		F	T	T	F	F	F	T
F	F	F		T	F	F	F	F	T	F

Now you're ready to move to the main operator, the →-operator outside the parentheses (flip back to Chapter 5 for information on how to determine the main operator). The two input values are in the columns under the *P* and the column under the other →-operator.

For example, in the first row, the value under the *P* is **T** and the value under the →-operator is **T**. And **T → T** evaluates to **T**, so write this directly under the main →-operator.

Finishing up the column gives you:

P	Q	P	→	(~	Q	→	(P	&	~	Q))
T	T	T	T	F	T	T	T	F	F	T
T	F	T	T	T	F	T	T	T	T	F
F	T	F	T	F	T	T	F	F	F	T
F	F	F	T	T	F	F	F	F	T	F

The column under the main operator should be the last column you fill in. If it isn't, you better erase (you were using pencil, weren't you?) and retrace your steps!

Reading a truth table

Circle the entire column under the main operator, so that the information jumps out at you when it's time to read the table. The column under the main operator is the most important column in the table. It tells you the truth value for each interpretation of the statement.

For example, if you want to know how the statement evaluates when *P* is false and *Q* is true, just check the third row of the table. The value in this row under the main operator is **T**, so when *P* is false and *Q* is true, the statement evaluates as true.

At this point, with great confidence you can go back to the original question:

What can you say about this statement: *P* → (~*Q* → (*P* & ~*Q*))? Justify your answer.

With your trusty truth table, you can tell your professor exactly what he or she wants to hear about the statement: that the statement is always true, regardless of the value of the constants P and Q.

And how do you justify this? You don't have to, because the truth table does it for you. As long as you filled it in correctly, the table tracks every possible interpretation of the statement. No other interpretations are possible, so your work is complete.

Putting Truth Tables to Work

After you know how to use truth tables, you can begin to understand SL at a whole new level. In this section, I show you how to tackle a few common problems about individual statements, pairs and sets of statements, and arguments. (In later chapters, I also show you how to tackle these same questions using a variety of different tools.)

Taking on tautologies and contradictions

Every statement in SL falls into one of these three categories: tautologies (true under every interpretation), contradictions (false under every interpretation), or contingent statements (either true or false depending upon the interpretation).

In the "Baby's First Truth Table" section, earlier, you see how you can use truth tables to sort out the truth value of a statement under every possible interpretation, which allows you to divide statements into three important categories:

- **Tautologies:** A *tautology* is a statement that is always true, regardless of the truth values of its constants. An example of a tautology is the statement $P \lor {\sim}P$. Because either P or ${\sim}P$ is true, at least one part of this or-statement is true, so the statement is always true.

- **Contradictions:** A *contradiction* is a statement that is always false, regardless of the truth values of its constants. An example of a contradiction is the statement $P \,\&\, {\sim}P$. Because either P or ${\sim}P$ is false, at least one part of this and-statement is false, so the statement is always false.

- **Contingent statements:** A *contingent statement* is a statement that is true under at least one interpretation and false under at least one interpretation. An example of a contingent statement is $P \rightarrow Q$. This statement is true when P is true and Q is true, but false when P is true and Q is false.

Don't make the mistake of thinking that every statement is either a tautology or a contradiction. Plenty of contingent statements don't fall into either category.

Truth tables are an ideal tool for deciding which category a particular statement falls into. For example, in the "Baby's First Truth Table" section earlier in this chapter, you used a truth table to show that the statement $P \to (\sim Q \to (P \& \sim Q))$ evaluates as true in every row of the truth table, so the statement is a tautology.

Similarly, if a statement evaluates as false in every row of a truth table, it's a contradiction. And, finally, if a statement evaluates as true in at least one row and false in at least one row, it's a contingent statement.

Judging semantic equivalence

When you look at single statements, you can use truth tables to evaluate them under all possible combinations of truth values for their constants. Now, you're going to take this process a step further and compare two statements at a time.

When two statements are *semantically equivalent*, they have the same truth value under all interpretations.

You already know a simple example of two statements that are semantically equivalent: P and $\sim\sim P$. When the value P is **T**, the value of $\sim\sim P$ is also **T**. Similarly, when the value of P is **F**, the value of $\sim\sim P$ is also **F**.

This example is easy to check because, with only one constant, there are only two possible interpretations. As you can probably imagine, the more constants you have, the hairier semantic equivalence becomes.

However, a truth table can help. For example, are the statements $P \to Q$ and $\sim P \lor Q$ semantically equivalent? You can find out by making a truth table for these two statements.

P	Q	P	→	Q	~	P	∨	Q
T	T							
T	F							
F	T							
F	F							

As I cover in the "Baby's First Truth Table" section earlier in the chapter, the first step for filling out the table is to copy the value of each constant into the proper columns:

P	Q	P	→	Q	~	P	V	Q
T	T	T		T		T		T
T	F	T		F		T		F
F	T	F		T		F		T
F	F	F		F		F		F

Second, handle any ~-operator that applies directly to a constant:

P	Q	P	→	Q	~	P	V	Q
T	T	T		T	F	T		T
T	F	T		F	F	T		F
F	T	F		T	T	F		T
F	F	F		F	T	F		F

Third, complete the evaluations on both statements separately, just as you would for a single statement:

P	Q	P	→	Q	~	P	V	Q
T	T	T	**T**	T	F	T	**T**	T
T	F	T	**F**	F	F	T	**F**	F
F	T	F	**T**	T	T	F	**T**	T
F	F	F	**T**	F	T	F	**T**	F

When two statements have the same truth value on every line of the truth table, they're semantically equivalent. Otherwise, they're not.

The table shows that in this case, the two statements are semantically equivalent. Chapter 8 shows how this important concept of semantic equivalence is applied to proofs in SL.

Staying consistent

If you can compare two statements at a time, why not more than two?

When a set of statements is *consistent,* at least one interpretation makes all of those statements true. When a set of statements is *inconsistent,* no interpretation makes all of them true.

For example, look at the following set of statements:

$P \lor \sim Q$

$P \rightarrow Q$

$P \leftrightarrow \sim Q$

Is it possible for all three of these statements to be true at the same time? That is, is there any combination of truth values for the constants P and Q that causes all three statements to be evaluated as true?

Again, a truth table is the tool of choice. This time, however, you put all three statements on the truth table. I've taken the liberty of filling in the correct information for the first statement. First, I copied the values of P and Q for each row in the correct columns. Next, I evaluated the value of ~Q. Finally, I evaluated the whole statement $P \lor \sim Q$, placing the value for each row under the main operator.

In three of the four rows, the statement evaluates as true. But when P is true and Q is false, the statement is false. Because you're looking for a row in which all three statements are true, you can rule out this row.

P	Q	P	∨	~	Q	P	→	Q	P	↔	~	Q
T	T	T	T	F	T							
T	F	T	T	T	F							
F	T	F	F	F	T	—	—	—	—	—	—	—
F	F	F	T	T	F							

When filling a truth table to test for consistency, move vertically as usual, but evaluate one statement at a time. When you find a row where the statement evaluates as false, draw a line all the way through it. Drawing a line through the row saves you a few steps by reminding you not to evaluate any other statement in that row.

Repeating this process for the next two statements gives the following result:

P	Q	P	∨	~	Q	P	→	Q	P	↔	~	Q
T	T	T	T	F	T	T	T	T	T	F	F	T
T	F	T	T	T	F	T	F	F	—	—	—	—
F	T	F	F	F	T	—	—	—	—	—	—	—
F	F	F	T	T	F	F	T	F	F	F	T	F

When every row of the truth table has at least one statement that evaluates as false, the statements are inconsistent. Otherwise, they're consistent.

In this case, you know that the three statements are inconsistent because under all interpretations, at least one statement is false.

Arguing with validity

As I discuss in Chapter 3, in a valid argument, when all the premises are true, the conclusion must also be true. Here is the same basic idea defined in terms of interpretations:

When an argument is *valid*, no interpretation exists for which all of the premises are true and the conclusion is false. When an argument is *invalid*, however, at least one interpretation exists for which all of its premises are true and the conclusion is false.

You can also use truth tables to decide whether an entire argument is valid. For example, here is an argument:

Premises:

$P \& Q$

$R \to \sim P$

Conclusion:

$\sim Q \leftrightarrow R$

In this case, the argument has three constants — P, Q, and R — so the truth table needs to have eight rows, because $2 \times 2 \times 2 = 8$ (refer to Table 6-1).

To set up a large truth table: Start in the rightmost constant column (R in this case) and write T, F, T, F, and so on, alternating every other row until you get to the end. Then move one column to the left and write T, T, F, F, and so on, alternating every two rows. Keep moving left and doubling, alternating next by fours, then by eights, and so forth until the table is complete.

Here is the truth table as it needs to be set up:

P	Q	R	P	&	Q	R	→	~	P	~	Q	↔	R
T	T	T											
T	T	F											
T	F	T											
T	F	F											
F	T	T											
F	T	F											
F	F	T											
F	F	F											

In the section "Staying consistent," I mention the advantage of evaluating an entire statement before moving on to the next statement. Here's the table after the first statement has been evaluated:

P	Q	R	P	&	Q	R	→	~	P	~	Q	↔	R
T	T	T	T	T	T								
T	T	F	T	T	T								
T	F	T	T	F	F	—	—	—	—	—	—	—	—
T	F	F	T	F	F	—	—	—	—	—	—	—	—
F	T	T	F	F	T	—	—	—	—	—	—	—	—
F	T	F	F	F	T	—	—	—	—	—	—	—	—
F	F	T	F	F	F	—	—	—	—	—	—	—	—
F	F	F	F	F	F	—	—	—	—	—	—	—	—

When you find a row where either a premise evaluates as false or the conclusion evaluates as true, draw a line all the way through it. Drawing a line through the row saves you a few steps by reminding you not to evaluate any other statement in this row.

The first column of this example is especially helpful because in six of the eight rows of the previous table, the first premise is false, which means you can rule out these six rows. Here is what the rest of the table looks like when completed:

P	Q	R	P	&	Q	R	→	~	P	~	Q	↔	R
T	T	T	T	T	T	T	F	F	T	—	—	—	—
T	T	F	T	T	T	F	T	F	T	F	T	T	F
T	F	T	T	F	F	F	—	—	—	—	—	—	—
T	F	F	T	F	F	F	—	—	—	—	—	—	—
F	T	T	F	F	T	—	—	—	—	—	—	—	—
F	T	F	F	F	T	—	—	—	—	—	—	—	—
F	F	T	F	F	F	—	—	—	—	—	—	—	—
F	F	F	F	F	F	—	—	—	—	—	—	—	—

When no row of the truth table contains all true premises and a false conclusion, the argument is valid; otherwise, it's invalid.

As you can see, the only row in the previous table that has all true premises also has a true conclusion, so this argument is valid.

Putting the Pieces Together

The previous sections of this chapter show how to use truth tables to test for a variety of logical conditions. Table 6-2 organizes this information.

Table 6-2	Truth Table Tests for a Variety of Logical Conditions	
Condition Being Tested	Number of Statements	Condition Verified When
Tautology	1	Statement is true in every row.
Contradiction	1	Statement is false in every row.
Contingent Statement	1	Statement is true in at least one row and false in at least one row.
Semantic Equivalence	2	Both statements have the same truth value in every row.
Semantic Inequivalence	2	The two statements have different values in at least one row.
Consistency	2 or more	All statements are true in at least one row.
Inconsistency	2 or more	All statements aren't true in any row.
Validity	2 or more	In every row where all premises are true, the conclusion is also true.
Invalidity	2 or more	All premises are true *and* the conclusion is false in at least one row.

If you have a sneaking suspicion that all of these concepts are somehow connected, you're right. Read on to see how they all fit together.

Connecting tautologies and contradictions

You can easily turn a tautology into a contradiction (or vice versa) by negating the whole statement with a ~-operator.

Recall from earlier that the statement

$$P \rightarrow (\sim Q \rightarrow (P \, \& \sim Q))$$

is a tautology. So its negation, the statement

$$\sim(P \rightarrow (\sim Q \rightarrow (P \, \& \sim Q)))$$

is a contradiction. To make sure this is so, here's the truth table for this new statement:

P	Q	~	(P	→	(~Q	→	(P	&	~Q)))
T	T	F	T	_T_	T	T	T	T	T
T	F	F	T	_T_	F	T	T	F	F
F	T	F	F	_T_	T	F	F	F	T
F	F	F	F	_T_	F	T	F	F	F

As you can see, the only thing that has changed is that the main operator of the statement — the only operator outside the parentheses — is now the ~-operator.

It should also be clear that you can turn a contradiction into a tautology in the same way. Thus, the statement

$$\sim(\sim(P \to (\sim Q \to (P \& \sim Q))))$$

is a tautology, which shows you that even though tautologies and contradictions are polar opposites, they're very closely linked.

Linking semantic equivalence with tautology

When you connect any two semantically equivalent statements with the ↔-operator, the resulting statement is a tautology.

As you saw earlier, the two statements

$P \to Q$ \qquad and \qquad $\sim P \lor Q$

are semantically equivalent. That is, no matter what truth values you choose for P and Q, the two statements have the same evaluation.

Now connect these two statement with a ↔-operator:

$$(P \rightarrow Q) \leftrightarrow (\sim P \vee Q)$$

The result is a tautology. If you doubt this result, look at the truth table for this statement:

P	Q	(P	→	Q)	↔	(~	P	∨	Q)
T	T	T	<u>T</u>	T	T	F	T	<u>T</u>	T
T	F	T	<u>F</u>	F	T	F	T	<u>F</u>	F
F	T	F	<u>T</u>	T	T	T	F	<u>T</u>	T
F	F	F	<u>T</u>	F	T	T	F	<u>T</u>	F

And, of course, you can turn this tautology into a contradiction by negating it as follows:

$$\sim((P \rightarrow Q) \leftrightarrow (\sim P \vee Q))$$

Linking inconsistency with contradiction

When you connect an inconsistent set of statements into a single statement by repeated use of the &-operator, the resulting statement is a contradiction.

As you saw earlier, the three statements

$P \vee \sim Q$

$P \rightarrow Q$

$P \leftrightarrow \sim Q$

are inconsistent. That is, under any interpretation, at least one of them is false.

Now connect these three statements with &-operators:

$$((P \vee \sim Q) \,\&\, (P \rightarrow Q)) \,\&\, (P \leftrightarrow \sim Q)$$

REMEMBER

When you use operators to connect more than two statements, you need to use extra parentheses so it's clear which operator is the main operator. I discuss this in more detail in Chapter 14.

The main operator is the second &-operator — the only operator outside the parentheses — but in any case, the result is a contradiction. To verify this result, you can use a table to evaluate the statement for all interpretations. First, evaluate everything inside the first set of parentheses:

P	Q	((P	∨	~	Q)	&	(P	→	Q))	&	(P	↔	~	Q)
T	T	T	T	F	T	T	T	T	T		T	F	F	T
T	F	T	T	T	F	F	T	F	F		T	T	T	F
F	T	F	F	F	T	F	F	T	T		F	T	F	T
F	F	F	T	T	F	T	F	T	F		F	F	T	F

Next, evaluate the entire statement:

P	Q	((P	∨	~	Q)	&	(P	→	Q))	&	(P	↔	~	Q)
T	T	T	T	F	T	T	T	T	T	F	T	F	F	T
T	F	T	T	T	F	F	T	F	F	F	T	T	T	F
F	T	F	F	F	T	F	F	T	T	F	F	T	F	T
F	F	F	T	T	F	T	F	T	F	F	F	F	T	F

As predicted, the statement evaluates as false on every line of the table, so it's a contradiction.

Linking validity with contradiction

As you may have guessed, argument validity can also be woven into this tapestry. For example, here's an old favorite example of a valid argument (yes, people who are into logic have old favorites):

Premises:

$P \to Q$

$Q \to R$

Conclusion:

$P \to R$

Because this argument's valid, you know it's impossible that both premises are true and the conclusion is false. In other words, if you filled in a truth table, none of the rows would look like this:

P	→	Q	Q	→	R	P	→	R
	T			T			F	

Similarly, if you negated the conclusion by using a ~-operator and then filled in another truth table, *none* of the rows would look like this:

P	→	Q	Q	→	R	~	(P	→	R)
	T			T				T	

But if *no* interpretation makes all of these statements true, you can consider this is an inconsistent set of statements.

Making even more connections

For those purists among you who just have to know how everything fits together:

✔ When you connect any two semantically inequivalent statements with the ↔ operator, the resulting statement is *not* a tautology — that is, it's either a contingent statement or a contradiction.

✔ When you negate the conclusion of an invalid argument, you get a consistent set of statements.

✔ When you connect an inconsistent set of statements into a single statement by repeated use of the &-operator, the resulting statement is *not* a contradiction — that is, it's either a contingent statement or a tautology.

When you negate the conclusion of a valid argument, you get a set of inconsistent statements. (It may seem a little backward that validity and inconsistency are linked, but that's the way it shakes out.) You can also turn this set of inconsistent statements into a contradiction by connecting them with the &-operator:

((P	→	Q)	&	(Q	→	R))	&	~	(P	→	R)
							T				

To turn a valid argument into a contradictory statement, connect all of the premises plus the negation of the conclusion by repeated use of the &-operator.

Chapter 7

Taking the Easy Way Out: Creating Quick Tables

In This Chapter

▶ Looking at the truth values of whole statements

▶ Understanding how to set up, fill in, and read a quick table

▶ Knowing what type of statements you're working with

---

*O*kay, if you've been reading this book for the last few chapters, you likely have the hang of the whole truth table thing, and you may be getting pretty good at them — maybe even too good for your own good.

For example, suppose early Monday morning, your professor walks into class with a cup of coffee, two doughnuts, and the morning paper. She gives you an in-class assignment to write up a truth table for the following statement:

$$P \rightarrow ((Q \& R) \vee (\sim P \& S))$$

Then, she sits down, opens her newspaper, and ignores everybody.

Ugh! With four constants, you're talking sixteen rows of evaluation purgatory. But, you're one of those troublemakers, so you go up to her desk and point out that the main operator of the statement is the →-operator, because it's the only operator outside the parentheses. She rattles her newspaper with annoyance.

You persist, carefully explaining to her your brand new insight: "Don't you see? All eight interpretations with *P* as false have to make the entire statement true. Isn't that right?" She takes a big bite of doughnut and uses this jelly-filled bite as an excuse not to say anything while she glares at you.

Finally, you get up the courage to ask her this: "So, how about if I just mark those eight rows as true, without going through all the steps?"

She snaps "No!", and you skulk back to your desk.

Like I said, you're a troublemaker. You're probably the type who reads (or maybe even writes) *For Dummies* books. Read on, MacDuff.

You're not too far off with your troublemaker thinking. There's a better way than plugging and chugging for problems like these (unless your professor is feeling particularly cruel) — the *quick table*! Unlike truth tables, which require you to evaluate a problem under *every* possible interpretation, quick tables use only one row to do the work of an entire truth table.

In this chapter, you see how quick tables save you time by working with statements as a *whole* instead of in *parts*, as truth tables do. I show you how to recognize the types of problems that are more easily solved with quick tables than with truth tables. And, I walk you through the strategies and methods that help you use quick tables to solve a variety of common problems.

Dumping the Truth Table for a New Friend: The Quick Table

Truth tables are ordered, precise, thorough — and tedious! They're tedious because you have to evaluate *every* possible interpretation in order to solve a problem.

When using truth tables, you start with the *parts* of a statement (the truth values of the constants) and finish with the *whole* statement (the value of the main operator). This method of starting with the parts and ending with the whole is both the strength and the weakness of truth tables. Because you must evaluate a statement under every possible interpretation, you're sure to cover all of the bases. For the same reason, however, a lot of your work is repetitive — in other words, tedious and boring!

But, as you saw happen with the troublemaking student in the chapter intro, a lot of these interpretations tend to be redundant. So, in many cases, you can eliminate bunches of them all at once. At the same time, however, you need to be careful to throw away only the wrong interpretations and hold onto the right ones. To make sure you're sure you're chucking the right interpretations, you need a system (and I just so happen to provide you with one in this chapter).

Just the opposite of truth tables, with quick tables, you start with the *whole* statement — the truth value of the main operator — and finish with the values of its *parts*, which are the values of the constants. The idea behind this method is to start out with just one truth value for a statement and, by making a few smart decisions, save a lot of time by eliminating repetitive work.

You can use quick tables in place of truth tables to test for any of the conditions discussed in Chapter 6 and listed there in Table 6-2.

Generally speaking, the following three types of problems are those types where you want to toss aside your old familiar truth tables in favor of quick tables:

- **Problems that your professor tells you to do using quick tables:** 'nuff said!

- **Problems with four or more constants:** Big tables mean big trouble. You're bound to make mistakes no matter how careful you are. In these cases, quick tables are virtually guaranteed to save you time.

- **Problems with statements that are "easy types":** Some types of statements are easy to crack with quick tables. I show you how to recognize them later in this chapter in the section "Working Smarter (Not Harder) with Quick Tables."

Outlining the Quick Table Process

In this section, I give you an overview of the three basic steps to using quick tables. An example is probably the best way to understand how to use these tables. So, I walk you through an example, and then I fill in more details in later sections.

Each of the three steps to using a quick table can be made easier if you know a few tricks about how to approach it. I provide those a bit later, but for now, just follow along and, if you have any questions, you can find answers later in the chapter.

Here's the example you'll follow through the next couple of sections: Suppose you want to know whether the following statements are consistent or inconsistent:

$$P \& Q \qquad Q \rightarrow R \qquad R \rightarrow P$$

Making a strategic assumption

All quick tables begin with a *strategic assumption*. In this example, you assume that all three statements are true, and you see where this leads.

Given this strategic assumption, every quick table can lead to two possible outcomes:

- ✔ Finding an interpretation under your assumption
- ✔ Disproving the assumption by showing that no such interpretation exists

Depending on the problem you're solving, each outcome leads to a different conclusion you can draw.

Think about the example problem this way: If the statements $P \& Q$, $Q \to R$, and $R \to P$ are consistent, the truth values of all three statements are **T** under at least one interpretation (check out Chapter 6 to review the definition of consistency).

So, a good strategy is to assume that the truth value of each statement is **T** and then see whether you can make this assumption work. For example, here's what your table would look like:

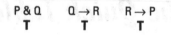

$$P \& Q \qquad Q \to R \qquad R \to P$$
$$\textbf{T} \qquad\qquad \textbf{T} \qquad\qquad \textbf{T}$$

Flip to the "Planning Your Strategy" section later in the chapter to see how to set up a quick table to solve every type of problem you've handled with truth tables. There, I provide you with a complete list of the strategic assumptions you should use for each type of problem.

Filling in a quick table

After the quick table is set up, you look for any further conclusions that can be drawn about the truth values of any parts of the statement. In the case of this section's ongoing example, because the statement $P \& Q$ is true, both sub-statements P and Q are also true:

$$\begin{array}{ccc} \text{P \& Q} & \text{Q} \rightarrow \text{R} & \text{R} \rightarrow \text{P} \\ \text{T T T} & \text{T} & \text{T} \end{array}$$

After you know that *P* and *Q* are both true, you can fill in this information everywhere that these constants appear, like this:

$$\begin{array}{ccc} \text{P \& Q} & \text{Q} \rightarrow \text{R} & \text{R} \rightarrow \text{P} \\ \text{T T T} & \text{T T} & \text{T T} \end{array}$$

Now look at the second statement, $Q \rightarrow R$. The entire statement is true and the first part is also true, so the second part has to be true as well. Therefore, *R* is true everywhere it appears. So, your table should now look like this:

$$\begin{array}{ccc} \text{P \& Q} & \text{Q} \rightarrow \text{R} & \text{R} \rightarrow \text{P} \\ \text{T T T} & \text{T T T} & \text{T T T} \end{array}$$

At this point, all of the constants are filled in, so you are ready to read your quick table.

Reading a quick table

When you've completely filled in a quick table, you have a *possible* interpretation, but you need to make sure that it really works

When you think you have an interpretation that works, check to make sure that

- ✔ Every constant has the same truth value everywhere it appears.
- ✔ Every evaluation is correct under that interpretation.

The example that I'm using passes both of these tests. Each of the three variables has the same truth value wherever it appears. (For example, the value of *P* is **T** throughout.) And, every evaluation is correct under this interpretation. (For example, the value of *P* & *Q* is **T**, which is correct.)

So you've found an interpretation that makes your original assumption correct, which means that all three statements are consistent.

Disproving the assumption

In the ongoing example from the preceding sections, the assumption led to an interpretation. But, as I state in the "Making a strategic assumption" section earlier in the chapter, this doesn't always happen. Sometimes, you may find that an assumption leads to an impossible situation.

For example, suppose you want to know whether the statement $(P \& Q) \& ((Q \leftrightarrow R) \& \sim P)$ is a contradiction — that is, whether its truth value is **F** under every interpretation.

As always, you begin with a strategic assumption. In this case, assume that the statement *isn't* a contradiction, so its truth value is **T** under at least one interpretation:

$(P \& Q) \& ((Q \leftrightarrow R) \& \sim P)$
T

As I discuss in Chapter 5, this statement's main operator — the only operator that appears outside of all parentheses — is the second &-operator. So this statement is of the form $x \& y$. Under the assumption that the whole statement is true, you can conclude that both sub-statements are also true, which would make your table look like this:

$(P \& Q) \& ((Q \leftrightarrow R) \& \sim P)$
T **T**   **T**

But, notice that both $P \& Q$ and $(Q \leftrightarrow R) \& \sim P$ are also of the form $x \& y$, which means that P, Q, $Q \leftrightarrow R$, and $\sim P$ are all true:

$(P \& Q) \& ((Q \leftrightarrow R) \& \sim P)$
ⓣT**T** T   **T**   T ⓣ

So far, so good. In just a couple of steps, you've made a lot of progress. However, here comes trouble: It seems that P and $\sim P$ are both true, which clearly can't be correct. This impossibility disproves the original assumption — which was that the statement *isn't* a contradiction. Therefore, the statement *is* a contradiction.

When you disprove an assumption, you need to be careful that you've truly ruled out all possible interpretations. Because the fact that you've ruled them out may not be self-evident from the finished quick table, some professors may require a brief explanation about how you arrived at your conclusion.

Here is the type of explanation that would do just fine for this example: "Assume the statement is *not* a contradiction. So, at least one interpretation makes the statement true. Then, (P & Q) and ((Q ↔ R) & ~P) are true. But then, both P and ~P are true, which is impossible, so the statement is a contradiction."

Planning Your Strategy

When testing for any of the conditions listed in Chapter 6, in Table 6-2, you begin with a strategic assumption and then look for an interpretation that fits. If you indeed find such an interpretation, you have one answer; if you find that no interpretation exists, you have another answer.

This section is a summary of the information you need to set up and read a quick table. For each case, I give you the assumption you need to start with, provide an example and your first step, and then tell you how to read both possible outcomes.

In each case, the strategic assumption I give you is the best way (and sometimes the only way) to test for the given condition using a quick table. That's because the assumption in each case allows you to draw a conclusion based on the existence (or non-existence) of a *single* interpretation — and quick tables are tailor-made to find a single interpretation if one exists.

Tautology

Strategic assumption: Try to show that the statement is not a tautology, so assume that the statement is false.

Example: Is $((P \to Q \to R)) \to ((P \, \& \, Q) \to R)$ a tautology?

First step:

$$(P \to Q \to R) \to ((P \, \& \, Q) \to R)$$
$$F$$

Outcomes:

✔ **If you find an interpretation under this assumption:** The statement is *not* a tautology — it's either a contradiction or a contingent statement.

✔ **If you disprove the assumption:** The statement is a tautology.

Contradiction

Strategic assumption: Try to show that the statement is not a contradiction, so assume that the statement is true.

Example: Is $((P \to Q \to R)) \to ((P \& Q) \to R)$ a contradiction?

First step:

$$(P \to Q \to R) \to ((P \& Q) \to R)$$
$$\mathsf{T}$$

Outcomes:

✔ **If you find an interpretation under this assumption:** The statement is *not* a contradiction — it's either a tautology or a contingent statement.

✔ **If you disprove the assumption:** The statement is a contradiction.

Contingent statement

Use the previous two tests for tautology and contradiction. If the statement *isn't* a tautology and *isn't* a contradiction, it must be a contingent statement.

Semantic equivalence and inequivalence

Strategic assumption: Try to show that the two statements are semantically inequivalent, so connect them using a $\leftrightarrow$-operator and assume that this new statement is false.

Example: Are $P \& (Q \lor R)$ and $(P \lor Q) \& (P \lor R)$ semantically equivalent statements?

First step:

$$(P \,\&\, (Q \lor R)) \leftrightarrow ((P \lor Q) \,\&\, (P \lor R))$$
$$F$$

Outcomes:

- ✔ **If you find an interpretation under the assumption:** The statements are semantically inequivalent.
- ✔ **If you disprove the assumption:** The statements are semantically equivalent.

Consistency and inconsistency

Strategic assumption: Try to show that the set of statements is consistent, so assume that all of the statements are true.

Example: Are the statements $P \,\&\, Q$, $\sim(\sim Q \lor R)$, and $\sim R \to \sim P$ consistent or inconsistent?

First step:

$$P \,\&\, Q \qquad \sim(\sim Q \lor R) \qquad \sim R \to \sim P$$
$$T \qquad\qquad T \qquad\qquad\qquad T$$

Outcomes:

- ✔ **If you find an interpretation under the assumption:** The set of statements is consistent.
- ✔ **If you disprove the assumption:** The set of statements is inconsistent.

Validity and invalidity

Strategic assumption: Try to show that the argument is invalid, so assume that all of the premises are true and the conclusion is false.

Example: Is this argument valid or invalid?

Premises:

$P \rightarrow Q$

$\sim(P \leftrightarrow R)$

Conclusion:

$\sim(\sim Q \& R)$

First step:

$$P \rightarrow Q \qquad \sim(P \leftrightarrow R) \qquad \sim(\sim Q \& R)$$
$$T \qquad\qquad\quad T \qquad\qquad\quad F$$

Outcomes:

⊩ **If you find an interpretation under the assumption:** The argument is invalid.

⊩ **If you disprove the assumption:** The argument is valid.

Working Smarter (Not Harder) with Quick Tables

Part of the trade-off in using quick tables is that you have to think about how to proceed instead of just writing out all of the possibilities. So, quick tables become a lot easier when you know what to look for. In this section, I show you how to use quick tables to your best advantage.

In Chapter 5, I discuss the eight basic forms of SL statements as a way to understand evaluation. These forms are even more helpful when you're using quick tables, so take a look at Table 5-1 if you need a quick refresher.

When you're working with quick tables, the truth value of each basic statement form becomes important. Two possible truth values (**T** and **F**) for each of the eight forms gives you 16 different possibilities. Some of these are easier to use in quick tables than others. I start with the easy ones.

Recognizing the six easiest types of statements to work with

Of the 16 types of SL statements (including truth values), 6 are easy to work with when using quick tables. With each of these types, the truth value of the statement's two sub-statements, x and y, are simple to figure out.

For example, suppose that you have a statement in the form $x \& y$ and you know its truth value is **T**. Remember that the only way a &-statement can be true is when both parts of it are true, so you know that the values of both x and y are also **T**.

Similarly, suppose you have a statement in the form $\sim(x \& y)$ and you know its truth value is **F**. In this case, it's easy to see that the value of $x \& y$ is **T**, which again means that the values of both x and y are **T**.

Figure 7-1 shows you the six easiest types of SL statements to work with. After you recognize these statements, you can often move very quickly through a quick table.

Starting with either:	Leads to:	Values of x and y:
$x \& y$ OR $\sim(x \& y)$ T F	$x \& y$ T T T	x is T and y is T
$x \lor y$ OR $\sim(x \lor y)$ F T	$x \lor y$ F F F	x is F and y is F
$x \to y$ OR $\sim(x \to y)$ F T	$x \to y$ T F F	x is T and y is F

Figure 7-1: The six easiest types of SL statements.

For example, suppose you want to find out whether the following is a valid or invalid argument:

Premises:

$\sim(P \to (Q \lor R))$

$\sim(P \& (Q \leftrightarrow \sim R))$

Conclusion:

(P & ~R)

The first step is always to pick the correct strategy. In this case, as shown in the "Planning Your Strategy" section earlier in the chapter, you assume the premises to be true and the conclusion to be false. (In other words, you assume that the argument is *invalid* and you look for an interpretation that fits this assumption.) Your table will look like this:

~(P → (Q ∨ R))	~(P & (Q ↔ ~R))	(P & ~R)
T	T	F

Notice that the first statement is in the form ~(x → y), with a truth value of **T**. Referring to Figure 7-1, you know that P is true and Q ∨ R is false, which means you can fill in your table like this:

~(P → (Q ∨ R))	~(P & (Q ↔ ~R))	(P & ~R)
T T F F	T	F

Now that you know the value of Q ∨ R is **F**, you can refer again to Figure 7-1 and see that Q and R are both false:

~(P → (Q ∨ R))	~(P & (Q ↔ ~R))	(P & ~R)
T T F F F F	T	F

In only three steps, you've figured out the truth values of all three constants, so you can fill these in like this:

~(P → (Q ∨ R))	~(P & (Q ↔ ~R))	(P & ~R)
T T F F F F	T T F F	T F F

Now, you need to finish filling in the table and check to see whether this interpretation works for every statement:

~(P → (Q ∨ R))	~(P & (Q ↔ ~R))	(P & ~R)
T T F F F F	T T F F F T F	(T F T) F

In this case, the second statement is correct. However, the third statement is incorrect: Both parts of the &-statement are true, so the value of the whole statement can't be **F**. This disproves the assumption that the argument is *invalid*, so you know that the argument is *valid*.

Working with the four not-so-easy statement types

Sometimes, you're going to get stuck with types of SL statements that aren't as easy to work with as those six I introduced in the previous section. This is especially true when you're testing for semantic equivalence because, as I discuss earlier in the chapter in "Planning Your Strategy," the strategy here is to join the two statements with a ↔-operator.

All four statement types that contain the ↔-operator give *two* possible sets of values for x and y, as shown in Figure 7-2.

Figure 7-2:
Four not-so-easy types of SL statements.

Starting with either:	Leads to:	Values of x and y:
x ↔ y OR ~(x ↔ y) **T** **F**	x ↔ y **T** T **T** **F** T **F**	EITHER x is **T** and y is **T** OR x is **F** and y is **F**
x ↔ y OR ~(x ↔ y) **F** **T**	x ↔ y **T** F **F** **F** F **T**	EITHER x is **T** and y is **F** OR x is **F** and y is **T**

Suppose you want to find out whether the statements ~($P \lor (Q \to R)$) and (($P \to R$) & Q) are semantically equivalent or inequivalent. The strategy here is to connect the two statements with a ↔-operator, and then you can assume that this new statement is false. (In other words, assume that the two statements are semantically inequivalent.) Your table would look like this:

$$\sim(P \lor (Q \to R)) \leftrightarrow ((P \to R) \& Q)$$
$$\mathbf{F}$$

As you can see from Figure 7-2, you can take the following two possible avenues for this statement:

$$\sim(P \lor (Q \to R)) \leftrightarrow ((P \to R) \& Q)$$

| T | F | F |
| F | F | T |

When looking at the first avenue, notice that the first part of the statement is one of the six easy types to work with, so you can fill in your table as shown here:

$$\sim(P \lor (Q \to R)) \leftrightarrow ((P \to R) \& Q)$$
T F F F   F   F

Furthermore, once you know that $Q \to R$ is false, you can conclude that Q is true and R is false. Now you can fill in your table like this:

$$\sim(P \lor (Q \to R)) \leftrightarrow ((P \to R) \& Q)$$
T F F T F F F   F

After you know the values of all three constants, you can fill in the rest of the table:

$$\sim(P \lor (Q \to R)) \leftrightarrow ((P \to R) \& Q)$$
T F F T F F F F (T) F (F T)

Under this interpretation, the two parts of the &-statement are true, but the statement itself is false, which is incorrect. So, the search for an interpretation that works goes on.

Now, you can try the second avenue:

$$\sim(P \lor (Q \to R)) \leftrightarrow ((P \to R) \& Q)$$
F   F   T

In this avenue, the second part of the statement is an easy type, so you can fill in the table as follows:

$$\sim(P \lor (Q \to R)) \leftrightarrow ((P \to R) \,\&\, Q)$$
$$\text{F} \qquad\qquad \text{F} \;\; \text{T} \;\; \text{T T}$$

Now, you know that Q is true. Also, the $\lor$-statement in the first part is true because its negation is false. So, you fill in the table, which should now look like this:

$$\sim(P \lor (Q \to R)) \leftrightarrow ((P \to R) \,\&\, Q)$$
$$\text{F} \;\; \text{T T} \qquad \text{F} \;\; \text{T} \;\; \text{T T}$$

At this point, you have no more solid conclusions to make. But, you're close to finished, and with a quick table you need to find only one interpretation that works. In this case, I suggest that you make a guess and see where it takes you.

Suppose, for example, that the value of P were **T**. Then to make the sub-statement $P \to R$ true, the value of R would also have to be **T**. So, it looks like you get a perfectly good interpretation of the statement when the values of all three constants are **T**. Fill in your table so that it looks like this:

$$\sim(P \lor (Q \to R)) \leftrightarrow ((P \to R) \,\&\, Q)$$
$$\text{F T}_\text{T} \text{ }_\text{T} \text{T T} \quad \text{F} \quad \text{T}_\text{T} \text{T}_\text{T} \text{T}$$

This checks out because, as I discussed in "Reading a quick table," every constant has the same truth value everywhere it appears, and the entire evaluation is correct under that interpretation. So, you've found an interpretation under your original assumption, which means that the two statements are semantically inequivalent.

You may be wondering what would have happened if you had guessed that the value of P were **F**. In this case, you would have found an alternative interpretation, with the truth value of R being **T**. But it doesn't matter — with quick tables, you need to find only one interpretation, and you're done.

Coping with the six difficult statement types

Six types of SL statements don't lend themselves very well to quick tables. Figure 7-3 shows why this is so.

Starting with either:	Leads to:	Values of x and y:
x & y OR ~(x & y) F T	x & y T F F F F T F F F	EITHER x is T and y is F OR x is F and y is T OR x is F and y is F
x ∨ y OR ~(x ∨ y) T F	x ∨ y T T T T T F F T T	EITHER x is T and y is T OR x is T and y is F OR x is F and y is T
x → y OR ~(x → y) T F	x → y T T T F T T F T F	EITHER x is T and y is T OR x is F and y is T OR x is F and y is F

Figure 7-3:
The six difficult types of SL statements.

As you can see, each of these types of statements leads to three possible avenues, and if you're using a quick table to solve this problem, you're in for a long, cumbersome search. Fortunately, you have other options. The following sections show you your options when you're facing three possible outcomes.

The first way: Use a truth table

One way to avoid the trouble of these types of statements is to fall back on truth tables. Truth tables are always an option (unless your professor forbids them), and they'll always give you the right answer. The downside of truth tables, as always, is that if a problem has a lot of different constants, you're in for a lot of work.

The second way: Use a quick table

The second way to solve a difficult problem is just to grit your teeth and try all three avenues with a quick table. After all, you figured out how to chase down two avenues in the previous example. And besides, you might get lucky and find an interpretation that works on the first try.

For example, suppose you want to find out whether the statement $(\sim(P \to Q)$ $\& \sim(R \lor S)) \to (Q \leftrightarrow R)$ is a contradiction. The strategy in this case is to make the following assumption:

$$(\sim(P \to Q) \& \sim(R \lor S)) \to (Q \leftrightarrow R)$$
$$\mathbf{T}$$

which leads to three possible avenues:

$$(\sim(P \to Q) \& \sim(R \lor S)) \to (Q \leftrightarrow R)$$

T	T	T
F	T	T
F	T	F

Fortunately, the first avenue leads someplace promising: Notice that the sub-statement $(\sim(P \to Q) \& \sim(R \lor S))$ is a &-statement whose value is **T**, which is one of the six easy types of statements. The table looks like this:

$$(\sim(P \to Q) \& \sim(R \lor S)) \to (Q \leftrightarrow R)$$
$$\text{T} \quad \text{T}\mathbf{T} \quad \text{T} \quad \text{T}$$

Even better, the two smaller statements are also easy types, leaving the table looking like this:

$$(\sim(P \to Q) \& \sim(R \lor S)) \to (Q \leftrightarrow R)$$
$$\text{TTF F TTFFF T T}$$

Now you can plug in the values of Q and R:

$$(\sim(P \to Q) \& \sim(R \lor S)) \to (Q \leftrightarrow R)$$
$$\text{TTF F TTFFF T FTF}$$

In just three steps, you found an interpretation that works under the assumption that the statement is true, so the statement is not a contradiction.

You won't always get this lucky, but you can see that in this case, a quick table still worked a lot faster than a truth table. Good luck!

The third way: Use a truth tree

In Chapter 8, I introduce *truth trees* as a perfect alternative for situations like these. When truth tables are too long and quick tables are too difficult, a truth tree may be your best friend.

Chapter 8

Truth Grows on Trees

....

....

*I*n Chapters 6 and 7, I show you how to use truth tables and quick tables to get information about sentential logic (SL) statements and sets of statements. In this chapter, I introduce you to the third (and my favorite!) method for solving logic problems: truth trees.

In the following sections, I show you how truth trees work by a process of decomposing SL statements into their sub-statements. Then I show you how to use truth trees to solve the same types of problems you've been handling with truth tables and quick tables.

Understanding How Truth Trees Work

Truth trees are a powerful tool by any standard. In fact, I think that they're the best tool in this book to solve nearly every type of problem you'll come across in logic. Why? Glad you asked.

First of all, truth trees are easy: They're easy to master and easy to use.

Truth trees combine the best features of both truth tables and quick tables, but without the baggage of either. For example, like truth tables, truth trees are a plug-and-chug method (though a couple of smart choices along the way can still help you out), but they're much shorter. And, like quick tables, truth trees avoid repetitive evaluation, but they never lead to trouble such as guesswork or problems with three possible avenues to eliminate.

Truth trees are a perfect way to solve SL problems of *any* size with maximum efficiency. They're also useful in quantifier logic (QL), the larger logical system I discuss in Part IV.

In Chapter 5, I discuss the eight basic forms of SL statements. These become important once again as you study truth trees, because truth trees handle each of these forms in a different way. (If you need a refresher on these eight forms, flip to Table 5-1.)

Decomposing SL statements

Truth trees work by *decomposing* statements — that is, breaking statements into smaller sub-statements.

For example, if you know that the statement $(P \lor Q) \& (Q \lor R)$ is true, you know that both the sub-statement $(P \lor Q)$ is true and the sub-statement $(Q \lor R)$ is true. In general, then, you can break any true statement of the form $x \& y$ into two true statements, x and y.

In some cases, decomposing a statement means breaking it off into two statements, where at least one of which is true. For example, if you know that the statement $(P \to Q) \lor (Q \leftrightarrow R)$ is true, you know that either the sub-statement $(P \to Q)$ or the sub-statement $(Q \leftrightarrow R)$ is true. In general, then, you can break any true statement of the form $x \lor y$ into two statements, x and y, where at least one of which is true.

Forms of true statements that lead directly to other true statements are called *single branching* statements. Forms that lead in two possible directions are called *double branching* statements. Double branching statements always lead to either one or two sub-statements per branch.

Figure 8-1 shows a list of all eight basic forms of SL statements with their decompositions.

For example, if you're decomposing the statement $(P \to Q) \lor (Q \to R)$, you begin by breaking it into two sub-statements as follows:

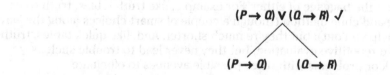

Notice that I checked the statement $(P \to Q) \lor (Q \to R)$ after I decomposed it into its two sub-statements. The checkmarks let you know that I'm finished with this statement — though now I need to deal with its two sub-statements.

Next, you break each sub-statement into smaller sub-statements, according to the rules provided in Figure 8-1.

Single Branching	Double Branching to Two Sub-Statements	Double Branching to One Sub-Statement
$x \& y$ x y	$x \leftrightarrow y$ $x \quad\quad \sim x$ $y \quad\quad \sim y$	$\sim(x \& y)$ $\sim x \quad\quad \sim y$
$\sim(x \lor y)$ $\sim x$ $\sim y$	$\sim(x \leftrightarrow y)$ $x \quad\quad \sim x$ $\sim y \quad\quad y$	$x \lor y$ $x \quad\quad y$
$\sim(x \to y)$ x $\sim y$		$x \to y$ $\sim x \quad\quad y$

Figure 8-1: The eight types of SL statements with decompositions.

You can see now how truth trees got their name. The final structure resembles an upside-down tree, like this:

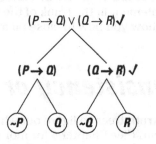

Here, I again checked the statements that I decomposed. Also note the circling convention. After you decompose a statement down to either a single constant or its negation, circling makes it easier to keep track of. In this case, I circled ~P, Q, ~Q, and R.

After you break down every statement in this way, the truth tree is complete. Each branch tells you something about one or more interpretations that would make the original statement true. To find these interpretations, trace from the beginning of the trunk all the way to the end of that branch and note all of the circled statements that you pass through along the way.

For example, tracing from the beginning of the trunk to the end of the first branch, the only circled statement you pass through is ~P. So, this branch tells you that *any* interpretation in which P is false will make the original statement true.

Solving problems with truth trees

You can use a truth tree to solve any problem that can be solved using a truth table or a quick table. As with these other tools, truth trees follow a step-by-step process. Here are the steps for a truth tree:

1. **Set up.** To set up a truth tree, construct its *trunk* according to the type of problem you're trying to solve.

   The trunk consists of the statement or statements you need to decompose.

2. **Fill in.** To fill in a truth tree, use the rules of decomposition listed in Figure 8-1 to create all of its *branches*.

3. **Read.** To read the completed truth tree, check to see which of the following two outcomes has occurred:

   • **At least one branch is left open:** At least one interpretation makes every statement in the trunk of the tree true.

   • **All of the branches are closed:** No interpretation makes every statement in the trunk of the tree true. (In the following section, I show you how to close off a branch.)

Showing Consistency or Inconsistency

You can use truth trees to figure out whether a set of statements is consistent or inconsistent. (See Chapter 6 for more on consistency.) For example, suppose you want to figure out whether the three statements $P \& \sim Q$, $Q \vee \sim R$, and $\sim P \to R$ are consistent or inconsistent.

To decide whether a set of statements is consistent (at least one interpretation makes all of those statements true) or inconsistent (no interpretation makes all of them true), construct a truth tree using that set of statements as its trunk. Here's your trunk:

$$P \& \sim Q$$
$$Q \vee \sim R$$
$$\sim P \to R$$

After you create your trunk, you can begin decomposing the first statement, $P \& \sim Q$. Here's what you get:

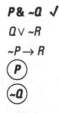

I checked the statement $P \& \sim Q$ after I decomposed it into its two substatements and circled the single constants — P and $\sim Q$.

The next statement is $Q \vee \sim R$, which decomposes along two separate branches as follows:

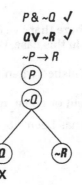

When tracing from the beginning of a trunk to the end of a branch would force you to pass through a pair of contradictory circled statements, close off that branch with an X.

In this case, tracing all the way from the beginning of the trunk to the end of the branch on the left forces you to pass through the three circled statements *P*, *~Q*, and *Q*. But *~Q* and *Q* are contradictory statements, so I've closed off that branch.

The reason for closing off this branch makes sense when you think about it. This branch tells you that any interpretation in which the statements *P*, *~Q*, and *Q* are true would make all three original statements true. But, *~Q* and *Q* can't both be true, so this branch provides *no* possible interpretations.

The final statement to decompose is *~P → R*. As you know from Figure 8-1, the branches for this statement will look like this:

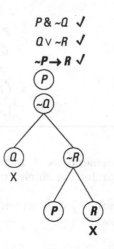

You can see again here that I've closed off a branch where a contradiction has been reached. In this case, the contradiction is *R* and *~R*.

Your truth tree is finished when one of the following occurs:

- ✔ Every statement or constant has been either checked or circled
- ✔ Every branch has been closed off

In this example, every item has been either checked or circled, so the tree is finished. After the tree is finished, check to see whether any branches are still open. The presence or absence of open branches on a finished truth tree allows you to determine whether the original set of statements is consistent or inconsistent. Follow these guidelines:

- ✔ If the finished truth tree has at least one open branch, the set of statements is *consistent*.
- ✔ If the finished truth tree has all closed branches, the set of statements is *inconsistent*.

As you can see, the example in this section still has one branch open, which means that an interpretation exists under which the three statements are all true. So, in this case, the set of statements is consistent.

If you want to know what this interpretation is, just trace from the trunk to the end of this branch. When you trace the length of the tree, you find that the circled items are *P, ~Q,* and *~R* and the final *P*. So, the only interpretation that makes the three original statements true is when the value of *P* is **T** and the values of both *Q* and *R* are **F**.

Testing for Validity or Invalidity

Truth trees are also handy when you want to determine an argument's validity or invalidity (see Chapter 6 for more on validity). For example, suppose you want to figure out whether the following argument is valid or invalid:

Premises:

   ~P ↔ Q

   ~(P ∨ R)

Conclusion:

   ~Q & ~R

To decide whether an argument is valid or invalid, construct a truth tree using the premises and the *negation* of the conclusion as its trunk.

Using the example I introduced at the beginning of this section, create a trunk that looks like this:

$$\sim P \leftrightarrow Q$$
$$\sim (P \lor R)$$
$$\sim (\sim Q \,\&\, \sim R)$$

TIP

You don't, however, have to start at the top of the tree. Decomposing statements in a different order is often helpful. Figure 8-1 divides the eight basic statement forms into three columns. This division can help you decide what order to decompose your statements. Whenever you have more than one statement in the trunk of your truth tree, decompose them in this order:

1. Single branching

2. Double branching with two sub-statements

3. Double branching with one sub-statement

Decomposing statements in this order makes sense. Whenever you can, choose a single branch path to keep your trees as small as possible. But, when you have no choice but to double branch, choose a double branch with two sub-statements. Adding two statements increases the chance that you'll be able to close off one of the branches.

In this example, only the second statement, $\sim (P \lor R)$, leads to a single branch. So, you decompose it first, like this:

$$\sim P \leftrightarrow Q$$
$$\sim (P \lor R) \;\checkmark$$
$$\sim (\sim Q \,\&\, \sim R)$$
$$\boxed{\sim P}$$
$$\boxed{\sim R}$$

Both of the remaining statements lead to double branches. But only the first statement, $(\sim P \leftrightarrow Q)$, decomposes to two sub-statements. So, you decompose that one next:

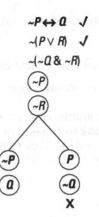

At this point, one branch is closed off: Tracing from the beginning of the trunk to the end of this branch forces you to pass through both ~P and P, which is a contradiction. The final step, decomposing ~(~Q & ~R), leads your truth tree to look like this:

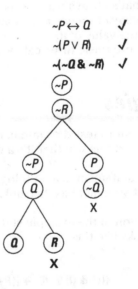

Notice that you only have to add the new decomposition to the open branch but not to the closed-off branch. Now every statement is either checked or circled, so the tree is complete.

When checking a truth tree for validity or invalidity the following guidelines apply:

  ✔ If the truth tree has at least one open branch, the argument is *invalid*.

  ✔ If the truth tree has all closed branches, the argument is *valid*.

In this section's example, one branch is still open, so the argument is invalid. Trace from the trunk to the end of this branch and you will find the circled statements ~*P*, *Q*, and ~*R*. This tells you that the *only* interpretation under which this argument is invalid is when the value of *P* and *R* are both **F** and the value of *Q* is **T**.

Separating Tautologies, Contradictions, and Contingent Statements

In Chapter 6, I show you that every statement in SL is categorized as a *tautology* (a statement that's always true), a *contradiction* (a statement that's always false), or a *contingent statement* (a statement that can be either true or false, depending on the value of its constants). You can use truth trees to separate SL statements into these three categories.

Tautologies

Suppose you want to test the statement $((P \& Q) \vee R) \rightarrow ((P \leftrightarrow Q) \vee (R \vee (P \& \sim Q)))$ to determine whether it's a tautology.

To show that a statement is a tautology, construct a truth tree using the *negation* of that statement as its trunk.

Using the negation of the example statement I just introduced, you can create a trunk that looks like this:

$$\sim(((P \& Q) \vee R) \rightarrow ((P \leftrightarrow Q) \vee (R \vee (P \& \sim Q))))$$

Even though this statement looks big and hairy, you know that it corresponds to one of the eight basic forms from Chapter 5 (listed in Table 5-1.) You just need to figure out which is the correct form. I go into this in Chapter 5, but a little refresher here wouldn't hurt.

The main operator is the first ~-operator — the only operator that is outside of all parentheses — so the statement is one of the four negative forms. And the scope of the →-operator covers the rest of the statement, so this statement is of the form ~$(x → y)$. As you can see in Figure 8-1, this form is single-branching, so your truth tree should now look like this:

$$\sim(((P \& Q) \lor R) \to ((P \leftrightarrow Q) \lor (R \lor (P \& \sim Q)))) \checkmark$$
$$(P \& Q) \lor R$$
$$\sim((P \leftrightarrow Q) \lor (R \lor (P \& \sim Q)))$$

Notice that I removed the outer parentheses from the sub-statement $((P \& Q) \lor R)$. This is a legitimate step, as I explain in Chapter 14.

Now, the last statement is of the form ~$(x \lor y)$, which is also single-branching. Therefore, according to the order in which you should decompose multiple statements (see the "Testing for Validity or Invalidity" section earlier in the chapter), that's the statement to work on next. Fill in your truth table as follows:

$$\sim(((P \& Q) \lor R) \to ((P \leftrightarrow Q) \lor (R \lor (P \& \sim Q))) \checkmark$$
$$(P \& Q) \lor R$$
$$\sim((P \leftrightarrow Q) \lor (R \lor (P \& \sim Q))) \checkmark$$
$$\sim(P \leftrightarrow Q)$$
$$\sim(R \lor (P \& \sim Q))$$

Again, the last statement is of the form ~$(x \lor y)$, so it's single branching, and therefore, next up:

$$\sim(((P \& Q) \lor R) \to ((P \leftrightarrow Q) \lor (R \lor (P \& \sim Q)))) \checkmark$$
$$(P \& Q) \lor R$$
$$\sim((P \leftrightarrow Q) \lor (R \lor (P \& \sim Q))) \checkmark$$
$$\sim(P \leftrightarrow Q)$$
$$\sim(R \lor (P \& \sim Q)) \checkmark$$
$$\boxed{\sim R}$$
$$\sim(P \& \sim Q)$$

Even though this example may look long, step back a minute and notice that you've already taken three steps without double-branching. Not having to double-branch saves you tons of work as you proceed because you have to worry about only one branch rather than two (or four, or eight!).

Eventually, though, you have to double-branch with this example. Because it branches to two sub-statements, start with the statement ~(P ↔ Q), according to the order of decomposition I provide in the "Testing for Validity or Invalidity" section earlier. Your truth tree should now look like this:

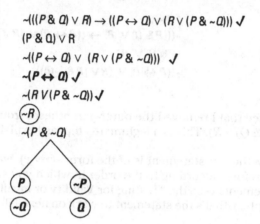

Now decompose (P & Q) ∨ R:

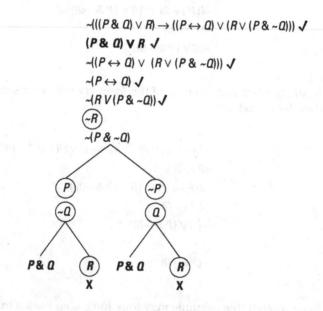

This step closes off two of the four branches. Now, P & Q is single branching, so decompose this statement on both remaining branches, like this:

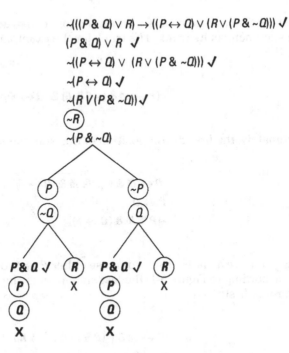

$$\sim(((P \& Q) \vee R) \rightarrow ((P \leftrightarrow Q) \vee (R \vee (P \& \sim Q)))) \checkmark$$
$$(P \& Q) \vee R \checkmark$$
$$\sim((P \leftrightarrow Q) \vee (R \vee (P \& \sim Q))) \checkmark$$
$$\sim(P \leftrightarrow Q) \checkmark$$
$$\sim(R \vee (P \& \sim Q)) \checkmark$$

You've now closed all remaining branches. Note that the statement $\sim(P \& \sim Q)$ is *not* checked. This makes no difference, though, because after every branch is closed off, the tree is finished.

When testing to determine whether a statement is a tautology, follow these guidelines:

✔ If the truth tree has at least one open branch, the statement is *not* a tautology — it's either a contradiction or a contingent statement. (To confirm or rule out contradiction as a possibility, you would need another tree, as I describe in the next section.)

✔ If the truth tree has no open branches, the statement is a tautology.

In this section's example, the tree shows you that the statement is a tautology.

Contradictions

Suppose you want to test the statement $(P \leftrightarrow Q) \& (\sim(P \& R) \& (Q \leftrightarrow R))$ to determine whether it's a contradiction.

To show that a statement is a contradiction, construct a truth tree using the statement as its trunk. The trunk for the example I just introduced looks like this:

$$(P \leftrightarrow Q) \& (\sim(P \& R) \& (Q \leftrightarrow R))$$

Fortunately, the first decomposition of this statement is single-branching:

$(P \leftrightarrow Q) \& (\sim(P \& R) \& (Q \leftrightarrow R))$ ✓
$P \leftrightarrow Q$
$\sim(P \& R) \& (Q \leftrightarrow R)$

Now you have a choice to decompose either $P \leftrightarrow Q$ or $\sim(P \& R) \& (Q \leftrightarrow R)$. But, according to Figure 8-1, the last statement is single-branching, so take that route first:

$(P \leftrightarrow Q) \& (\sim(P \& R) \& (Q \leftrightarrow R))$ ✓
$P \leftrightarrow Q$
$\sim(P \& R) \& (Q \leftrightarrow R)$ ✓
$\sim(P \& R)$
$Q \leftrightarrow R$

Because you've run out of single-branching statements, it's time to double-branch, starting with statements that produce two sub-statements. Begin with $P \leftrightarrow Q$:

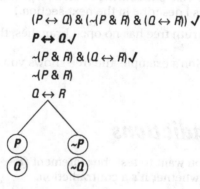

And now decompose $Q \leftrightarrow R$:

$$(P \leftrightarrow Q) \,\&\, (\neg(P \,\&\, R) \,\&\, (Q \leftrightarrow R)) \checkmark$$
$$P \leftrightarrow Q \checkmark$$
$$\neg(P \,\&\, R) \,\&\, (Q \leftrightarrow R) \checkmark$$
$$\neg(P \,\&\, R)$$
$$Q \leftrightarrow R \checkmark$$

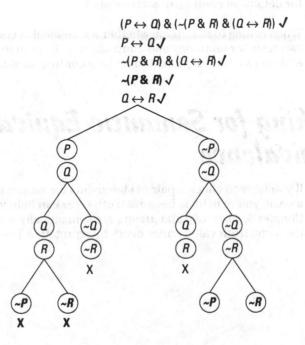

This step closes off two of the four branches. Your last step is to decompose $\neg(P \,\&\, R)$ like this:

$$(P \leftrightarrow Q) \,\&\, (\neg(P \,\&\, R) \,\&\, (Q \leftrightarrow R)) \checkmark$$
$$P \leftrightarrow Q \checkmark$$
$$\neg(P \,\&\, R) \,\&\, (Q \leftrightarrow R) \checkmark$$
$$\neg(P \,\&\, R) \checkmark$$
$$Q \leftrightarrow R \checkmark$$

Because every statement is either checked or circled, the tree is now complete.

When testing to determine whether a statement is a contradiction, apply these guidelines:

- ✔ If the truth tree has at least one open branch, the statement is *not* a contradiction — it's either a tautology or a contingent statement. (To confirm or rule out tautology as a possibility, you would need another tree, as I describe in the previous section.)

- ✔ If the truth tree has no open branches, the statement is a contradiction.

In this example, two branches remain open, so the statement isn't a contradiction.

Even though two branches remain open in this tree, only one interpretation makes the original statement true. Trace from the trunk to the end of either open branch to see that this interpretation is that *P, Q,* and *R* are all false.

Contingent statements

Checking to see whether a statement is contingent is, as always, just a matter of ruling out that it's either a tautology or a contradiction. (Flip to Chapter 6 for details on contingent statements.)

When testing to determine whether a statement is contingent, use the previous two tests for tautology and contradiction. If the statement *isn't* a tautology and *isn't* a contradiction, it must be a contingent statement.

Checking for Semantic Equivalence or Inequivalence

If you have to check a pair of statements for semantic equivalence or inequivalence, you're in luck, because truth trees can help you out. (As I explain in Chapter 6, when two statements are semantically equivalent, they both have the same truth value under every interpretation.)

REMEMBER

To decide whether a pair of statements is semantically equivalent or inequivalent, you have to construct two truth trees:

✔ One tree using the first statement and the negation of the second statement as its trunk

✔ The other tree using the negation of the first statement and the second statement as its trunk

Suppose you want to find out whether the statements $\sim P \to (Q \to \sim R)$ and $\sim(P \lor \sim Q) \to \sim R$ are semantically equivalent or inequivalent. In this case, you need to make two trees, with trunks as follows:

Tree #1:

$\sim P \to (Q \to \sim R)$

$\sim(\sim(P \lor \sim Q) \to \sim R)$

Tree #2:

$\sim(\sim P \to (Q \to \sim R))$

$\sim(P \lor \sim Q) \to \sim R$

Starting with Tree #1, the second statement is of the single-branching form $\sim(x \to y)$, so decompose this statement first:

$\sim P \to (Q \to \sim R)$

$\sim(\sim(P \lor \sim Q) \to \sim R)$ ✓

$\sim(P \lor \sim Q)$

$\boxed{R}$

The statement $\sim(P \lor \sim Q)$ is also single-branching, so work on it next:

$\sim P \to (Q \to \sim R)$

$\sim(\sim(P \lor \sim Q) \to \sim R)$ ✓

$\sim(P \lor \sim Q)$ ✓

$\boxed{R}$

$\boxed{\sim P}$

$\boxed{Q}$

Next, move on to the first statement, since you skipped it initially:

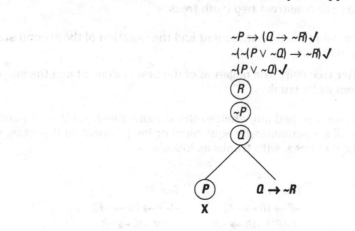

$$\sim P \rightarrow (Q \rightarrow \sim R)\ \checkmark$$
$$\sim(\sim(P \vee \sim Q) \rightarrow \sim R)\ \checkmark$$
$$\sim(P \vee \sim Q)\ \checkmark$$

This step closes off one branch. Now, decomposing $Q \rightarrow \sim R$, you get:

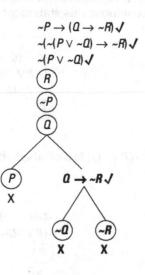

$$\sim P \rightarrow (Q \rightarrow \sim R)\ \checkmark$$
$$\sim(\sim(P \vee \sim Q) \rightarrow \sim R)\ \checkmark$$
$$\sim(P \vee \sim Q)\ \checkmark$$

Tree #1 is now complete.

When testing a pair of statements to determine whether they're semantically equivalent or inequivalent, follow these guidelines:

✔ If *either* truth tree has at least one open branch, the statements are *semantically inequivalent*.

✔ If *both* truth trees have all closed branches, the statements are *semantically equivalent*.

If the first tree has at least one open branch, the statements are *semantically inequivalent,* which means you can skip the second tree.

Because the first tree in the example has all closed branches, you need to move on to the next tree. In this case, the first statement is single-branching, so decompose it first as follows:

$$\sim(\sim P \rightarrow (Q \rightarrow \sim R)) \; \checkmark$$
$$\sim(P \vee \sim Q) \rightarrow \sim R$$
$$\boxed{\sim P}$$
$$\sim(Q \rightarrow \sim R)$$

The statement $\sim(Q \rightarrow \sim R)$ is also single-branching, so work on it next:

$$\sim(\sim P \rightarrow (Q \rightarrow \sim R)) \; \checkmark$$
$$\sim(P \vee \sim Q) \rightarrow \sim R$$
$$\boxed{\sim P}$$
$$\sim(Q \rightarrow \sim R) \; \checkmark$$
$$\boxed{Q}$$
$$\boxed{R}$$

The remaining statement, $\sim(P \vee \sim Q) \rightarrow \sim R$, is double-branching, so decompose it next, like this:

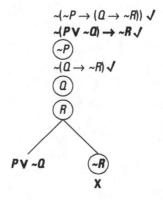

This step closes off one branch. Finally, decomposing $P \lor \sim Q$ gives you:

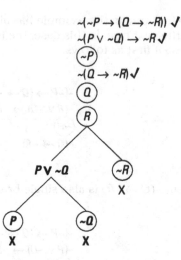

All of the branches on each tree are closed off, so you know that the statements are semantically equivalent.

Part III
Proofs, Syntax, and Semantics in SL

The 5th Wave   By Rich Tennant

"He ran some truth tables, threw in a few theorems, and before I knew it I was buying rust proofing."

In this part . . .

Proofs are the very heart of logic. For some students, though, writing proofs in logic is where they *lose* heart. But fear not! Proofs aren't as difficult as you've been led to believe — if you take the right approach.

In this part, you can expect to master SL proofs. Chapter 9 shows you what a proof looks like and how to construct one. You also discover the first eight rules of inference, which make up the set of implications rules. In Chapter 10, I discuss the remaining ten rules of inference, which are considered the set of equivalence rules. Chapters 9 and 10 focus on direct proof methods, while in Chapter 11, I introduce you to two new proof methods: conditional proof and indirect proof. In Chapter 12, I show you how and when to use all of these tools as I discuss proof strategies.

You also get the big picture of SL. In Chapter 13, you discover why the five SL operators are sufficient to produce any logical function in SL. In Chapter 14, I discuss a variety of topics related to the syntax and semantics of SL. Here, I show you how to decide whether a string of symbols in SL is also a well-formed formula. And finally, I give you a taste of Boolean algebra.

Chapter 9

What Have You Got to Prove?

. . . .

In This Chapter
▶ Introducing formal direct proof
▶ Constructing proofs using the rules of inferences

. . . .

*Y*ou may already have some experience writing proofs, those tricky problems that were an important part of high school geometry. In geometry proofs, you started with a set of simple axioms (also called postulates), such as "All right angles are equal," and figured out how to build toward ever more complex statements called theorems.

Computer programming, in which you use simple statements to create complex software, also resembles the proof method. This idea of complexity arising out of simplicity is also common to proofs in sentential logic (SL).

In a sense, constructing a proof is like building a bridge from one side of a river to another. The starting point is the set of premises you're given, and the end point is the conclusion you're trying to reach. And the pieces you use to build the bridge are the *rules of inference,* a set of 18 ways to turn old statements into new ones.

In this chapter, I introduce you to the first eight rules of inference — the set of *implications rules.* And along the way, you find out a bit about what proofs look like.

Even though these rules are clear and unambiguous, it's not always obvious how to use them in a particular case. There's an art to writing proofs that can make the process interesting and satisfying when you get it right, but frustrating when you don't. The good news is that a bunch of great tricks exist and, once you know them, you'll have a lot of ammo to use when the going gets tough.

Bridging the Premise-Conclusion Divide

A valid argument is like a bridge — it gives you a way from here (the premises) to there (the conclusion), even if rough waters are rushing below. Proofs give you a way to build such a bridge so that you know it is safe to travel from one side to the other.

For example, take the following argument.

Premises:

$P \rightarrow Q$

P

Conclusion:

Q

Because proofs concentrate so heavily on arguments, in this chapter I introduce a new space-saving convention for writing arguments. Using this convention, you can write the preceding argument as

$P \rightarrow Q, P : Q$

As you can see, a comma separates the premises $P \rightarrow Q$ and P, and a colon separates the conclusion Q. When you say that an argument is valid, you're saying in effect that your bridge is safe; that is, if you start on the left side (with the premises being true), you're safe to cross over to the right side (the conclusion is also true).

As a short example from arithmetic, look at this simple addition problem:

$2 + 3 = 5$

This equation is correct, so if you add 1 to both sides, you also get a correct equation:

$2 + 3 + 1 = 5 + 1$

Checking that the equation is still correct is easy because both sides add up to six.

Now, try the same thing using variables:

$a = b$

Suppose you add a third variable to both sides:

$a + c = b + c$

Is this equation correct? Well, you can't say for sure because you don't know what the variables stand for. But, even without knowing what they stand for, you can say that if the first equation is correct, then the second equation *must* also be correct.

So, the following is a valid statement:

$a = b \rightarrow a + c = b + c$

With this information, you can build the following argument:

Premises:

$a = b$

$a = b \rightarrow a + c = b + c$

Conclusion:

$a + c = b + c$

After you know that the left side of a valid argument is true, you know that you're safe to cross over to the right side — in other words, that the right side is also true. And it doesn't matter what numbers you fill in. As long as you start out with a true statement and keep the general *form* of the argument, you always end up with a true statement.

Using Eight Implication Rules in SL

Proofs in SL work very much like proofs in arithmetic. The only difference is that instead of using arithmetic symbols, they use the SL operators you've come to know and love (refer to Chapter 4 for more on these fun-loving operators).

SL provides eight *implication rules*, which are rules that allow you to build a bridge to get from here to there. In other words, you start with one or more true statements and end up with another one. Most of these rules are simple — they may seem even trivial. But, this simplicity is their power because it allows you to handle much more complex ideas with great confidence.

In this section, I introduce these eight rules and show you how to use them to write proofs in SL. Along the way, I give you a few tricks for making this task a bit easier. (See Chapter 12 for more in-depth coverage of proof strategies.) I also provide you with a number of ways to remember these rules for when exam time rolls around.

At the end of the day, you'll need to memorize these rules. But, work with them for a while and you may find that you remember most of them without too much effort.

The → rules: Modus Ponens and Modus Tollens

In Chapter 3, I compare if-statements to a slippery slide: If the first part of the statement is true, then the second part *must* be true in order for the statement to be true. Both implication rules in this section — *Modus Ponens* (**MP**) and *Modus Tollens* (**MT**) — take advantage of this idea in different ways.

Modus Ponens (MP)

MP takes advantage of the slippery slide idea directly: It tells you "If you know that a statement of the form $x \rightarrow y$ is true, and you also know that the x part is true, then you can conclude that the y part is also true." In other words, once you step on the top of the slide at x, you have no choice but to end up at y.

MP: $x \rightarrow y, x : y$

MP says that *any* argument of this form is valid. Here it is in its simplest form:

$P \rightarrow Q, P : Q.$

But, by the same rule, these similar arguments are all valid as well:

$\sim P \rightarrow \sim Q, \sim P : \sim Q$

$(P \& Q) \rightarrow (R \& S), P \& Q : R \& S$

$(P \leftrightarrow \sim(Q \& \sim R)) \rightarrow (\sim S \vee (R \rightarrow P)), (P \leftrightarrow \sim(Q \& \sim R)) : (\sim S \vee (R \rightarrow P))$

A typical proof has a bunch of numbered rows. The top rows are the premises, the last row is the conclusion, and the rows in-between are the intermediate steps to logically connect them. Each row has a line number and a statement, followed by a justification for that statement (which includes the rule and the row numbers it affects).

For example, here's a proof for the argument $(P \& Q) \to R, (P \& Q) : R$:

1.	$(P \& Q) \to R$	**P**
2.	$(P \& Q)$	**P**
3.	R	1, 2 **MP**

As you can see, this proof didn't require any intermediate steps. The premises (**P**) led immediately to the conclusion, with the justification being the rule **MP** applied to lines 1 and 2.

Now try to prove that this slightly more complicated argument is valid:

$$P \to Q, Q \to R, R \to S, P : S$$

The first step to any proof is always the same: Copy down all of the premises with line numbers and justifications. Here's what your first step should look like:

1.	$P \to Q$	**P**
2.	$Q \to R$	**P**
3.	$R \to S$	**P**
4.	P	**P**

After you've copied your premises, look for a step that you can take. In this case, you can use **MP** with lines 1 and 4:

5.	Q	1, 4 **MP**

MP allows you to derive a new statement, Q, which, you can use as part of your next step. This time, you can use **MP** with lines 2 and 5:

6.	R	2, 5 **MP**

Again, you derived a new statement, R, which you can use for your next step. The last step practically writes itself:

7.	S	3, 6 **MP**

You know you're finished when the conclusion of the argument appears. In this case, S is the conclusion you were looking to justify, so you're done.

Modus Tollens (MT)

MT uses the slippery slide idea in a different way from how MP uses it: It tells you "If you know that a statement of the form $x \rightarrow y$ is true, and you also know that the y part is false, you can conclude that the x part is also false." In other words, if you *don't* end up on the bottom of the slide at y, then you *didn't* step onto it in the first place at x.

MT: $x \rightarrow y, \sim y : \sim x$

As with MP (shown in the previous section), MT and all of the other rules of inference can be generalized. Thus, a few easily provable valid arguments in this form are as follows:

$$P \rightarrow Q, \sim Q : \sim P$$

$$(P \mathbin{\&} Q) \rightarrow R, \sim R : \sim(P \mathbin{\&} Q)$$

$$(P \lor Q) \rightarrow (R \leftrightarrow S), \sim(R \leftrightarrow S) : \sim(P \lor Q)$$

Knowing this rule, you can now prove the validity of this argument:

$$P \rightarrow Q, \sim P \rightarrow R, \sim Q : R$$

As always, begin by copying the premises:

1.	$P \rightarrow Q$	P
2.	$\sim P \rightarrow R$	P
3.	$\sim Q$	P

So far, I've given you two rules to work with — MP and MT. You'll need both in this proof.

In most proofs, short statements are easier to work with than long ones.

The shortest statement here is $\sim Q$, so a good plan is to hunt around for a way to use it. You don't have to hunt too far, because you can use MT as follows:

4.	$\sim P$	1, 3 MT

Now you can use MP:

5.	R	2, 4 MP

Again, the proof is finished when the conclusion appears.

The & rules: Conjunction and Simplification

The two & rules are related by a common factor: &-statements. *Simplification* (**Simp**) is useful for breaking down &-statements at the beginning of a proof and *Conjunction* (**Conj**) for building them up at the end.

Conjunction (Conj)

Conj says "If you know two things separately, then you also know both of them together."

Conj: $x, y : x \& y$

Conj is a pretty straightforward rule: If you have two true statements, x and y, you can conclude that the statement $x \& y$ is also true.

Try this proof on for size:

$$P \rightarrow Q, R \rightarrow S, P, \sim S : (P \& \sim S) \& (Q \& \sim R)$$

First, copy the premises:

1.	$P \rightarrow Q$	**P**
2.	$R \rightarrow S$	**P**
3.	P	**P**
4.	$\sim S$	**P**

Writing a proof can be a little like a treasure hunt: Try to find your way to the next clue by whatever means you can. For example, study the conclusion so you know where you're trying to go. Then look at the premises and see what will help you get there.

Generally speaking, long conclusions make for easier proofs than short ones. The strategy is to try to build a long conclusion one sub-statement at a time.

In this proof, you want to build the sub-statements $(P \& \sim S)$ and $(Q \& \sim R)$. The operators in these sub-statements are &-operators, which clues you in that **Conj** can help you build them. In fact, you can get one of them in one step:

5.	$P \& \sim S$	3, 4 **Conj**

Look for statements that have constants in common, and see whether you can combine them using the rules of inference.

Looking at the example, you can see that lines 1 and 3 have constants in common, and so do lines 2 and 4:

6.	Q	1, 3 **MP**
7.	~R	2, 4 **MT**

Now you can combine these two statements using **Conj**:

8.	Q & ~R	6, 7 **Conj**

This gives you the other sub-statement of the conclusion. The only thing left to do is to build the conclusion using the pieces you've gathered:

9.	$(P$ & ~$S)$ & $(Q$ & ~$R)$	5, 8 **Conj**

This was a long, nine-line proof, but when you work it piece by piece, it falls together.

Now that you've walked through this proof, copy down the argument, close the book, and see whether you can do it by yourself. You may find a few sticking points along the way, but it's better to discover them here than on an exam!

Simplification (Simp)

Simp tells you, "If you have two things together, than you also have either of those things separately."

Simp: x & $y : x$

x & $y : y$

Simp is sort of the flip side of **Conj**. But, instead of starting with the pieces and building to the whole, you start with the whole and reduce to either of the pieces.

Try writing the proof for this argument:

$P \rightarrow Q, R \rightarrow S, P$ & ~$S : Q$ & ~R

As always, copy down your premises, like so:

1.	$P \rightarrow Q$	**P**
2.	$R \rightarrow S$	**P**
3.	$P \,\&\, {\sim}S$	**P**

Use **Simp** early in a proof to *unpack* &-statements — that is, turn long &-statements into shorter statements. This gives you easier statements to work with, which you can use to build your conclusion.

Line 3 is your only opportunity to use this rule:

4.	P	3, **Simp**
5.	${\sim}S$	3, **Simp**

You may be noticing a certain similarity among the proofs in this chapter. If you are, that's good! These next two steps should look pretty familiar by now:

6.	Q	1, 4 **MP**
7.	${\sim}R$	2, 5 **MT**

All that remains is to put the pieces together, like this:

8.	$Q \,\&\, {\sim}R$	6, 7 **Conj**

The ∨ rules: Addition and Disjunctive Syllogism

As **Simp** is related to **Conj**, Disjunctive Syllogism (**DS**) is similarly related to Addition (**Add**). Both rules work with ∨-statements. **DS** breaks them down and **Add** builds them up.

Addition (Add)

Add tells you, "If you know x, then you can conclude either x or y."

Add: $x : x \lor y$

At first glance, this rule can seem downright weird. You may be asking "If I'm starting only with x, where did the y come from?" Believe it or not, the beauty of **Add** is the seemingly magical appearance of the y out of thin air.

Remember, to build a ∨-statement that's true, you only need one part of it to be true (see Chapter 4). The other part can be anything you like.

Try out this proof:

$$Q \to S, Q: ((P \leftrightarrow \sim Q) \leftrightarrow R) \lor ((P \lor S) \& (Q \lor R))$$

This proof sure looks screwy. There isn't much to work with — just two premises:

1.	$Q \to S$	**P**
2.	Q	**P**

The first step almost suggests itself:

3.	S	1, 2 **MP**

When the conclusion of an argument is a ∨-statement, you only need to build one of the two sub-statements, and then you can use **Add** to tack on the rest.

The key thing to realize here is that you have two choices: You can either prove the first part of the conclusion — $((P \leftrightarrow \sim Q) \leftrightarrow R)$ — or the second part — $((P \lor S) \& (Q \lor R))$. Place your bets on the second part, because I haven't yet given you any rules to handle ↔-statements.

A proof is like a bridge: The bigger the bridge, the more likely it was built from two sides rather than just one. So, for tougher proofs like this one, write the conclusion you're trying to reach at the bottom of the page and work your way up.

In this case, the conclusion will look like this:

6.	$(P \lor S) \& (Q \lor R)$	
7.	$((P \leftrightarrow \sim Q) \leftrightarrow R) \lor ((P \lor S) \& (Q \lor R))$	6 **Add**

With this setup, basically I'm saying "I'm not sure yet how I got here, but the last step was to tack on that hairy ↔-statement stuff using **Add**." (By the way, don't worry too much about statement numbers while you're working backwards — it just so happens that I have ESP.)

Now, look at line 6. Again, while working backwards, the question to ask is "How could I end up here?" This time, you notice that $(P \lor S) \& (Q \lor R)$ is an &-statement. And one way to build an &-statement is by tacking together the two parts of the statement with **Conj:**

4.	$P \lor S$	
5.	$Q \lor R$	
6.	$(P \lor S) \,\&\, (Q \lor R)$	4, 5 **Conj**
7.	$((P \leftrightarrow \sim Q) \leftrightarrow R) \lor ((P \lor S) \,\&\, (Q \lor R))$	6 **Add**

Again, you're not sure exactly *how* you got here, but if you can find a way to build the statements $P \lor S$ and $Q \lor R$, the rest follows.

Now, it's easier to see how the magic happened in the first place, because building $P \lor S$ and $Q \lor R$ aren't really so difficult. Look back up to lines 2 and 3 and use **Add** to bridge the gap. Here's how the whole proof looks from start to finish:

1.	$Q \rightarrow S$	**P**
2.	Q	**P**
3.	S	1, 2 **MP**
4.	$P \lor S$	2 **Add**
5.	$Q \lor R$	3 **Add**
6.	$((P \lor S) \,\&\, (Q \lor R))$	4, 5 **Conj**
7.	$((P \leftrightarrow \sim Q) \leftrightarrow R) \lor ((P \lor S) \,\&\, (Q \lor R))$	6 **Add**

Disjunctive Syllogism (DS)

DS says "If you have two options and you can eliminate one of them, you can be sure of the one that's left."

DS: $x \lor y, \sim x : y$

$x \lor y, \sim y : x$

DS is related to **Add** in this way: **DS** breaks down $\lor$-statements, and **Add** builds them up.

See what you can do with this argument:

$P \rightarrow \sim Q, P \lor R, Q \lor S, \sim R : \sim P \lor S$

First you copy your premises:

1.	$P \rightarrow \sim Q$	**P**
2.	$P \lor R$	**P**
3.	$Q \lor S$	**P**
4.	$\sim R$	**P**

In this case, the simplest statement is ~R. Line 2 also contains the constant R. These two statements allow you to use **DS** immediately:

5.	P		2, 4 **DS**

Now that you have the statement P, the next statement follows easily:

6.	~Q		1, 5 **MP**

Now you have another opportunity to use **DS**:

7.	S		3, 6 **DS**

And, finally, don't miss the opportunity to use **Add**:

8.	~P ∨ S		7 **Add**

The Double → Rules: Hypothetical Syllogism and Constructive Dilemma

Hypothetical Syllogism (**HS**) and *Constructive Dilemma* (**CD**) allow you to draw conclusions when you start out with two →-statements. You won't use them as often as the other six rules I mention in this chapter, but you'll still need them from time to time.

Hypothetical Syllogism (HS)

HS makes sense when you look at it. It tells you "If you know that x leads to y and that y leads to z, then x leads to z."

HS: $x \rightarrow y, y \rightarrow z : x \rightarrow z$

Note that **HS** is the first rule so far that contains no single constants. It neither breaks down statements nor builds them up.

Here's an example for you to try:

$$P \rightarrow Q, Q \rightarrow R, R \rightarrow S, {\sim}S : {\sim}P \, \& \, (P \rightarrow S)$$

As usual, write out your premises:

1.	$P \rightarrow Q$		P
2.	$Q \rightarrow R$		P
3.	$R \rightarrow S$		P
4.	~S		P

Looking at the conclusion, you need to get two pieces — $\sim\!P$ and $P \rightarrow S$ — and then put them together using **Conj**. You can get them in either order. I start by using **HS**:

5.	$P \rightarrow R$	1, 2 **HS**
6.	$P \rightarrow S$	3, 5 **HS**

That gives you the first piece. But now, the second piece isn't difficult to find:

7.	$\sim\!P$	4, 6 **MT**

Then, just put them together using **Conj**:

8.	$\sim\!P \,\&\, (P \rightarrow S)$	5, 7 **Conj**

Constructive Dilemma (CD)

CD is less intuitive than the other rules in this chapter until you really think about it: In plain English, it says "Suppose you know that you have either w or x. And you also know that w gives you y and that x gives you z. Well, then you have either y or z."

CD: $w \lor x,\ w \rightarrow y,\ x \rightarrow z : y \lor z$

This rule is also the only one that uses three statements to produce one, which allows for few opportunities to use **CD**. However, when such an opportunity arises, it's usually hanging there like neon sign. In other words, it's super easy to find.

In this example, I give you six premises in a desperate attempt to camouflage this opportunity:

$P \rightarrow Q,\ Q \rightarrow R,\ S \rightarrow T,\ U \rightarrow V,\ S \lor U,\ \sim\!R : (\sim\!P \,\&\, \sim\!Q) \,\&\, (T \lor V)$

So, first write out your premises:

1.	$P \rightarrow Q$	**P**
2.	$Q \rightarrow R$	**P**
3.	$S \rightarrow T$	**P**
4.	$U \rightarrow V$	**P**
5.	$S \lor U$	**P**
6.	$\sim\!R$	**P**

At first glance, this example is just a huge mess. One thing to notice, though, is that the conclusion is an &-statement. This means that if you can get both parts of it — $\sim\!P \,\&\, \sim\!Q$ and $T \lor V$ — you can use **Conj** to get the whole thing.

First, use **CD** to get $T \lor V$:

7.	$T \lor V$	3, 4, 5 **CD**

Now, how do you get $\sim P \,\&\, \sim Q$? Of course, just get $\sim P$ and $\sim Q$. Not so bad, right? Here's what these steps look like:

8.	$\sim Q$	2, 6 **MT**
9.	$\sim P$	1, 8 **MT**
10.	$\sim P \,\&\, \sim Q$	8, 9 **Conj**

To complete the proof, just put them together with **Conj**, like so:

11.	$(\sim P \,\&\, \sim Q) \,\&\, (T \lor V)$	7, 10 **Conj**

Chapter 10

Equal Opportunities: Putting Equivalence Rules to Work

*I*f you like implication rules (see Chapter 9), you're going to adore the other ten rules of inference — the *equivalence rules* — that I go over in this chapter. Why? Let me count the ways.

First of all, these rules will dazzle and astound you (and numerous other admirers) with the ease that they provide you in working logic problems. For example, just try proving the validity of the following argument using only implication rules:

~(P & Q), P : ~Q

Alas, you can't. But, lucky for you, with the new and improved equivalence rules in this chapter, problems like these are just momentary distractions on an otherwise cloudless day. And there's even more good news: Equivalence rules are generally easier and more flexible to use within proofs for several important reasons, which are also covered here.

In this chapter, you discover how to apply ten important equivalence rules, get tips on when and how to use them, and continue to sharpen your skill at proving the validity of arguments. The proofs in this chapter also make full use of the implication rules.

Distinguishing Implications and Equivalences

The implication rules that I discuss in Chapter 9 have several key limitations that the equivalence rules in this chapter *don't* have. In this way, the equivalence rules are more flexible and generally more useful than the implication rules. Read on to find out the differences between these two sets of rules.

Thinking of equivalences as ambidextrous

One of the most important differences between equivalences and implications is how they work: Equivalences work in both directions whereas implications work only in one direction.

For example, when you know that $x \rightarrow y$ and x are both true, Modus Ponens (**MP**) tells you that y is also true (see Chapter 9). However, reversing that deduction gets you into trouble because knowing that y is true is certainly not enough information to decide that $x \rightarrow y$ and x are both true.

Luckily, this limitation doesn't apply to the ten equivalence rules. Equivalence rules tell you that two statements are *interchangeable*: anywhere you can use one statement, you can use the other, and vice versa.

Applying equivalences to part of the whole

Another difference between equivalences and implications is that equivalences can be applied to part of a statement, but implications can't. Here's an example of an obvious mistake when working with implications:

1. $(P \& Q) \rightarrow R$ **P**
2. P 1, **Simp** (WRONG!)

Remember that the Simplification rule (**Simp**) states that $x \& y : x$. So, the mistake here is thinking that you can apply the implication rule **Simp** to part of line 1 ($P \& Q$). Equivalence rules, however, aren't bound by such restrictions.

Discovering the Ten Valid Equivalences

Let me guess, you're just chomping at the bit to get to know these ten equivalence rules, which you'll need to memorize, by the way. Well, here they are, complete with examples.

One bit of notation you'll need to know is the double-colon symbol ::. When placed between two statements, this symbol means that the two statements are equivalent — that is, you can substitute one for the other whenever needed.

Double Negation (DN)

DN is simple. It tells you "If *x* is true, then *not not x* is also true."

DN: $x :: \sim\sim x$

If you read Chapter 9, you can probably work through the following proof in your sleep, without using any equivalence rules:

$\sim P \to Q, \sim Q : P$

1.	$\sim P \to Q$	**P**
2.	$\sim Q$	**P**
3.	P	1, 2 **MT**

However, every time you negate a negation — for example, when you negate $\sim P$ and change it into P — you're technically required to follow these steps:

1. Negate $\sim P$ by changing it into $\sim\sim P$.

2. Use Double Negation (**DN**) to change $\sim\sim P$ into P.

So, the proof at the beginning of this section is technically missing a step — see the following version:

$\sim P \to Q, \sim Q : P$

1.	$\sim P \to Q$	**P**
2.	$\sim Q$	**P**
3.	$\sim(\sim P)$	1, 2 **MT**
4.	P	3 **DN**

Check to see if your teacher is a stickler for this technicality. If so, be careful not to assume a **DN** without putting it explicitly in your proof.

I'm going out on a brittle limb here: Throughout this chapter, I use **DN** frequently without referencing it. My philosophy is that you don't need to get all bent out of shape every time you flip from negative to positive. So sue me if you must, but I *don't* know *nothing* about double negation.

Contraposition (Contra)

Chapter 4 tells you that a statement of the form $x \rightarrow y$ and its contrapositive ($\sim y \rightarrow \sim x$) always have the same truth value. Chapter 6 clues you in that two statements that always have the same truth value are semantically equivalent. These two facts put together give you **Contra**.

Contra: $x \rightarrow y :: \sim y \rightarrow \sim x$

Contra is related to Modus Tollens (**MT**), which I introduced in Chapter 9. Each of these rules is saying: "When you start out knowing that the statement $x \rightarrow y$ is true, then the fact that $\sim y$ is true leads quickly to $\sim x$."

An easy way to think of **Contra** is: *Reverse* and *negate both*. That is, you can *reverse* the two parts of a $\rightarrow$-statement as long as you *negate both* parts of it. For example, take a look at this proof:

$$P \rightarrow Q, \sim P \rightarrow R : \sim R \rightarrow Q$$

| 1. | $P \rightarrow Q$ | **P** |
| 2. | $\sim P \rightarrow R$ | **P** |

This proof gives you two opportunities to use **Contra**:

| 3. | $\sim Q \rightarrow \sim P$ | 1 **Contra** |
| 4. | $\sim R \rightarrow P$ | 2 **Contra** |

Now, you can complete the proof in one step using **HS**:

| 5. | $\sim R \rightarrow Q$ | 1, 4 **HS** |

By the way, notice that in this proof you didn't use line 3 and, in fact, you can cross it out now if you like. (Your professor probably won't take off points for showing unnecessary steps, but if you're unsure, then cross it out.)

You don't have to use every statement or even every premise (though most of the time you will need them all). But, it can be useful to write down whatever statements you can derive and then see if you need them. You'll learn more about this strategy of jotting down the easy stuff in Chapter 12.

Implication (Impl)

The rational behind **Impl** is simple: When you know that the statement $x \to y$ is true, you know that either x is false (that is, $\sim x$ is true) or that y is true. In other words, you know that the statement $\sim x \lor y$ is true.

Impl: $x \to y :: \sim x \lor y$

What you probably didn't know, though, is that this rule works in reverse (as do all valid equivalences). So, whenever you're given $x \lor y$, you can change it to $\sim x \to y$. Professors who are sticklers may insist that you first change $x \lor y$ to $\sim\sim x \lor y$ using **DN** and then change this to $\sim x \to y$ using **Impl**.

An easy way to think of **Impl** is: *Change* and *negate the first*. For example, you change a $\to$-statement to a $\lor$-statement (or vice versa), as long as you *negate* the *first part* of the statement. Check out this proof:

$$P \to Q, P \lor R : Q \lor R$$

1.	$P \to Q$	**P**
2.	$P \lor R$	**P**

As you can see, this is a tough little proof because you don't have much to go on and it doesn't have any single-constant statements.

Impl links every $\lor$-statement with a $\to$-statement. Also, **Contra** gives you two versions of every $\to$-statement. That's three forms to work with.

For example, $P \to Q$ has two equivalent forms: $\sim P \lor Q$ by **Impl** and $\sim Q \to \sim P$ by **Contra**. Writing both out involves no cost or obligation and seeing them may suggest something, so be sure to write out all of your steps as you see them. Here they are:

3.	$\sim P \lor Q$	1 **Impl**
4.	$\sim Q \to \sim P$	1 **Contra**

The same is true of $P \lor R$:

5.	$\sim P \to R$	2 **Impl**
6.	$\sim R \to P$	2 **Contra**

Now look at what you're trying to prove: $Q \lor R$. With **Impl**, this is the same as $\sim Q \to R$. Aha! So, here are the next steps:

7.	$\sim Q \to R$	4, 5 **HS**
8.	$Q \lor R$	7 **Impl**

You didn't need lines 3 and 6, but they're free, so who cares? (If the answer to this rhetorical question is "My professor cares — he takes off points for extra steps," then go back and cross out as needed.)

When in doubt on a proof, write down as many statements as you can. It can't hurt and often helps. I call this strategy *kitchen-sinking* because you're throwing everything but the kitchen sink at the problem. I discuss it and other proof strategies in greater detail in Chapter 12.

Exportation (Exp)

Exp isn't intuitively obvious until you put it into words (which I do for you momentarily), and then its meaning jumps right out at you.

Exp: $x \to (y \to z) :: (x \& y) \to z$

To understand this rule, think of an example where $x \to (y \to z)$ could be the form of an English statement. Here's one possibility:

If I go to work today, *then if* I see my boss, *then* I'll ask for a raise.

Now think of a similar example with $(x \& y) \to z$:

If I go to work today *and* I see my boss, *then* I'll ask for a raise.

These two statements essentially mean the same thing, which should tell you that the two statement forms they're based on are semantically equivalent.

Don't mix up the parentheses! **Exp** tells you nothing about the statement $(x \to y) \to z$ or the statement $x \& (y \to z)$.

Try this proof on for size:

$(P \& Q) \to R, \sim R \lor S, P : Q \to S$

1.	$(P \& Q) \to R$	P
2.	$\sim R \lor S$	P
3.	P	P

Statement 1 is in one of the two forms that **Exp** works with, so try it:

    4. $P \to (Q \to R)$ 1, **Exp**

Now, another avenue opens up:

    5. $Q \to R$ 3, 4 **MP**

Okay, so you have $Q \to R$, but now you need $Q \to S$. If only you could get your hot little hands on $R \to S$, which would allow you to build the bridge you need using **HS**. Consider these steps:

    6. $R \to S$ 2 **Impl**

    7. $Q \to S$ 5, 6 **HS**

Commutation (Comm)

You may remember the commutative property from arithmetic as the rule telling you that order doesn't affect addition and multiplication — for example, $2 + 3 = 3 + 2$ and $5 \times 7 = 7 \times 5$.

In SL, **Comm** tells you that order doesn't affect operations with either the &-operator or the ∨-operator. Thus, **Comm** has two versions:

Comm: $x \& y :: y \& x$

           $x \lor y :: y \lor x$

Like **DN**, **Comm** may seem so obvious that you think it almost doesn't need to be mentioned. But, unlike **DN**, I think it's worth mentioning when you use it in a proof, and your professor probably will, too.

Here's an example of a proof where **Comm** comes in handy:

    $P \& (\sim Q \to R) : (R \lor Q) \& P$

    1. $P \& (\sim Q \to R)$ **P**

    2. P 1 **Simp**

    3. $\sim Q \to R$ 1 **Simp**

    4. $Q \lor R$ 3 **Impl**

Here comes your **Comm** opportunity — don't miss it!

    5. $R \lor Q$ 4 **Comm**

    6. $(R \lor Q) \& P$ 5 **Conj**

Association (Assoc)

Assoc tells you that for statements with all &-operators or with all ∨-operators, you can move the parentheses around freely.

Assoc: $(x \mathbin{\&} y) \mathbin{\&} z :: x \mathbin{\&} (y \mathbin{\&} z)$

$(x \lor y) \lor z :: x \lor (y \lor z)$

Like **Comm**, **Assoc** also has counterparts in arithmetic. For example, $(3 + 4) + 5 = 3 + (4 + 5)$

Assoc and **Comm** can be powerful tools when used together. By using just these two rules, you can rearrange, in any way you like, any statement that consists of all &-statements or all ∨-statements. Just be careful that you do it step by step.

Here's a proof to try out:

$(P \rightarrow Q) \lor R : Q \lor (R \lor \sim P)$

As always, write the premise first:

   1. $(P \rightarrow Q) \lor R$ **P**

Notice that the conclusion has only ∨-statements. So, if you can find a way to write the premise with only ∨-statements, you'll be able to finish the proof using only **Comm** and **Assoc**. Here's the next step.

   2. $(\sim P \lor Q) \lor R$ 1 **Impl**

Notice that I used **Impl** on only *part* of the premise.

At this point, the strategy is just to rearrange the constants in line 2 to make this statement look like the conclusion. Because the first constant in the conclusion is Q, the next step will get the Q in its proper position:

   3. $(Q \lor \sim P) \lor R$ 2 **Comm**

Notice that, again, I applied the equivalence rule to part of the statement. Next, I want to move the parentheses to the right using **Assoc**:

   4. $Q \lor (\sim P \lor R)$ 3 **Assoc**

All that's left to do now is switch around $\sim R$ and P:

   5. $Q \lor (R \lor \sim P)$ 4. **Comm**

This particular proof also works in reverse — that is, you can use the conclusion to prove the premise. This is true of *any* proof that has only one premise and uses only equivalence rules. This also tells you that the premise and conclusion are semantically equivalent.

Distribution (Dist)

As with **Exp**, **Dist** is confusing until you put it into words, and then it makes sense.

Dist: $x \& (y \lor z) :: (x \& y) \lor (x \& z)$

$x \lor (y \& z) :: (x \lor y) \& (x \lor z)$

This rule, too, has an analogous form in arithmetic. For example:

$$2 \times (3 + 5) = (2 \times 3) + (2 \times 5)$$

For this reason, multiplication *distributes* over addition. Note that the reverse isn't true:

$$2 + (3 \times 5) \neq (2 + 3) \times (2 + 5)$$

In SL, however, the &-operator and the ∨-operator distribute over each other. Here's how it works. I start with a statement in English that matches the form $x \& (y \lor z)$:

I have a pet *and* it's *either* a cat *or* a dog.

And here's the parallel statement for $(x \& y) \lor (x \& z)$:

Either I have a pet *and* it's a cat *or* I have a pet *and* it's a dog.

These two statements mean the same thing, which should help you understand why **Dist** works.

Similarly, here's how the ∨-operator distributes over the &-operator. This time, I start with a statement that matches the form $x \lor (y \& z)$:

I have to take *either* organic chemistry *or both* botany *and* zoology.

The parallel statement for $(x \lor y) \& (x \lor z)$ is a little awkward to translate, but here it is:

I have to take *either* organic chemistry *or* botany, *and* I also have to take *either* organic chemistry *or* zoology.

Spend a moment comparing this statement with the last to convince yourself that they mean the same thing.

Here's an example of a proof to get you started:

$Q \vee R, \sim(P \& Q), P : R$

1.	$Q \vee R$	**P**
2	$\sim(P \& Q)$	**P**
3.	P	**P**

You could apply **Impl** and **Contra** to statement 1, which wouldn't hurt, and I recommend this tactic when you're hunting around for ideas. However, I take this in another direction to show you how to use **Dist**:

4.	$P \& (Q \vee R)$	1, 3 **Conj**
5.	$(P \& Q) \vee (P \& R)$	4 **Dist**

Now, the wheels are turning. What next?

When you get stuck in the middle of a proof, look at the premises that you haven't used to get ideas.

Notice that the second premise is really just the negation of the first part of statement 5, which allows you to use **DS**:

6.	$P \& R$	2, 5 **DS**
7.	R	6 **Simp**

This is a tough little proof, and you may be wondering what you would have done if you hadn't known to use **Conj** to derive statement 4. In the next section, I show you how to tackle this same proof in an entirely different way.

DeMorgan's Theorem (DeM)

Like **Dist** and **Exp**, **DeM** also becomes clearer with an example using statements in English.

DeM: $\sim(x \& y) :: \sim x \vee \sim y$

$\sim(x \vee y) :: \sim x \& \sim y$

Here's a statement in English that matches the form ~(*x* & *y*). Notice that this statement is the *not . . . both* situation I discuss in Chapter 4:

It's *not true* that I'm *both* rich *and* famous.

And here's the parallel statement matching the form ~*x* ∨ ~*y*:

Either I'm *not* rich *or* I'm *not* famous.

As you can see, these two statements mean the same thing, which gives you an intuitive grasp on why the corresponding forms are equivalent.

Here's a statement in English that matches the form ~(*x* ∨ *y*). Notice that this statement is the *neither . . . nor* situation, which I discuss in Chapter 4:

Jack is *neither* a doctor *nor* a lawyer.

And here is the parallel statement matching ~*x* & ~*y*:

Jack *isn't* a doctor *and* he *isn't* a lawyer.

Use **DeM** to change statements of the forms ~(*x* & *y*) and ~(*x* ∨ *y*) into forms that are easier to work with.

Here's the argument I used in the preceding section:

Q ∨ *R*, ~(*P* & *Q*), *P* : *R*

1.	*Q* ∨ *R*	**P**
2	~(*P* & *Q*)	**P**
3.	*P*	**P**

This time, instead of setting up to use **Dist**, I start by applying **DeM** to statement 2:

4.	~*P* ∨ ~*Q*	2 **DeM**

Because this statement is so much easier to work with, you're more likely to see this step:

5.	~*Q*	3, 4 **DS**

The proof now practically finishes itself:

6.	*R*	1, 5 **DS**

You can almost always find more than one way to get to where you're going within a proof. If your first attempt doesn't work, try another route.

Tautology (Taut)

Taut is the simplest rule in the chapter.

Taut: $x \,\&\, x :: x$

$x \lor x :: x$

In fact, an "Uh-duh!" may be in order here. So why bother?

To be honest, you don't really need **Taut** in the &-statement case. You can change $x \,\&\, x$ to x by using **Simp** and can change x to $x \,\&\, x$ by using **Conj**. Similarly, in the ∨-statement case, you can change x to $x \lor x$ by using **Add**. However, if you're stuck with $x \lor x$ and need to get x, **Taut** is the way to go.

Even so, I can think of only one way you'd be likely to run into this kind of situation, and it's kind of cute:

$P \to {\sim}P : {\sim}P$

I can hear you already: "Is this proof even possible?" Yes, it is. Take a look:

1.	$P \to {\sim}P$	**P**
2.	${\sim}P \lor {\sim}P$	1 **Impl**
3.	${\sim}P$	2 **Taut**

So, given that the statement $P \to {\sim}P$ is true, you can prove that ${\sim}P$ is also true.

Equivalence (Equiv)

You may think that I lost track of ↔-statements. In fact, you may notice that of the 18 rules of inference I discuss in this chapter and Chapter 9, **Equiv** is the only one that even includes a ↔-statement.

Equiv: $x \leftrightarrow y :: (x \to y) \,\&\, (y \to x)$

$x \leftrightarrow y :: (x \,\&\, y) \lor ({\sim}x \,\&\, {\sim}y)$

Now, the truth can be told: I've been ignoring these statements because they're so darned ornery to work with in proofs.

One reason for the relative orneriness of ↔-statements is that their truth tables are perfectly symmetrical. In contrast to the tables for the other three binary operators, the table for the ↔-operator divides evenly into two true statements and two false ones. This symmetry is very pretty visually, but it isn't much help when you want to narrow a field down to *one* possibility.

So, when an argument contains a ↔-operator, you're going to have to get rid of it, and the sooner, the better. And you're in luck because the two **Equiv** rules can help you do just that.

The first of the two **Equiv** forms exploits the idea that in a ↔-statement, the arrow goes in both directions. It just pulls the statement apart into two →-statements and connects them with a &-operator. The second form of **Equiv** exploits the idea that both parts of a ↔-statement have the same truth value, which means that either x and y are both true or both false.

Take a look at this argument:

$$P \leftrightarrow (Q \& R), Q : P \vee {\sim}R$$

1.	$P \leftrightarrow (Q \& R)$	**P**
2.	Q	**P**

When you have a ↔-statement as a premise, automatically write both **Equiv** forms in your proof:

3.	$(P \rightarrow (Q \& R)) \& ((Q \& R) \rightarrow P)$	1 **Equiv**
4.	$(P \& (Q \& R)) \vee ({\sim}P \& {\sim}(Q \& R))$	1 **Equiv**

While you're at it, use **Simp** to unpack the &-statement version of **Equiv**:

5.	$P \rightarrow (Q \& R)$	3 **Simp**
6.	$(Q \& R) \rightarrow P$	3 **Simp**

Voila! You've written four statements automatically. At this point, look to see what you have. In this case, line 6 looks like an **Exp** waiting to happen:

7.	$Q \rightarrow (R \rightarrow P)$	6 **Exp**

From here, notice that you haven't even touched line 2 yet, and it all falls into place like so:

8.	$R \rightarrow P$	2, 7 **MP**
9.	${\sim}R \vee P$	8 **Impl**
10.	$P \vee {\sim}R$	9 **Comm**

In this case, you didn't need the second **Equiv** form. But, check out this proof where you do need it:

$$P \leftrightarrow Q, R \rightarrow (P \vee Q) : R \rightarrow (P \& Q)$$

This is the most advanced proof in this chapter.

Study all the steps to make sure you understand them, then try to reproduce it with the book closed. If you master this, you'll have a good grasp on some of the most difficult of the 18 rules of inference.

As always, the premises come first:

1.	$P \leftrightarrow Q$	**P**
2.	$R \rightarrow (P \vee Q)$	**P**

And here are the four statements you can write out without much thought, all from the first premise:

3.	$(P \rightarrow Q) \& (Q \rightarrow P)$	1 **Equiv**
4.	$(P \& Q) \vee (\sim P \& \sim Q)$	1 **Equiv**
5.	$P \rightarrow Q$	3 **Simp**
6.	$Q \rightarrow P$	3 **Simp**

Now, notice that line 2 is $R \rightarrow (P \vee Q)$, and the conclusion is $R \rightarrow (P \& Q)$. So, if you could find a way to derive $(P \vee Q) \rightarrow (P \& Q)$, then the implication rule **HS** would give you the conclusion.

In the end, though, only line 4 proves useful and allows for the following fancy footwork:

7.	$\sim(P \& Q) \rightarrow (\sim P \& \sim Q)$	4 **Impl**
8.	$\sim(\sim P \& \sim Q) \rightarrow (P \& Q)$	7 **Contra**
9.	$(P \vee Q) \rightarrow (P \& Q)$	8 **DeM**
10.	$R \rightarrow (P \& Q)$	2, 6 **HS**

Chapter 11

Big Assumptions with Conditional and Indirect Proofs

*H*ave you ever seen one of those cheesy infomercials where the announcer keeps asking, "*Now* how much would you pay?" as he throws in a bunch of extra stuff you don't want with your $19.95 oven-safe goggles — the cheese grater, ice crusher, and the potato peeler that writes under water. This chapter is going to be a bit like those commercials, but with one key difference: I'm going to throw in all sorts of extra stuff that you really *do* want.

In this chapter, I introduce two new forms of proof — *conditional proof* and *indirect proof* — completely free of charge. Unlike the *direct proofs* I discuss in Chapters 9 and 10, conditional proof and indirect proof involve an *assumption,* which is an additional premise that you don't *know* is true but you *assume* is true.

Also at no further cost to you, in this chapter, I show you not only how but also when to use each of these methods.

Conditional proof almost always makes a proof easier, but you can't always use it. However, when the conclusion to an argument is a →-statement, conditional proof is usually the way to go. On the other hand, indirect proof, also called *proof by contradiction,* is an industrial strength method that works for every proof. However, even though it's sometimes the *only* method that will work, it isn't necessarily always the *easiest* way, so use it sparingly.

Conditioning Your Premises with Conditional Proof

So, you've been a diligent student — you've practiced writing proofs until you can do it in your sleep. You've even memorized every implication and equivalence rule from Chapters 9 and 10. Congratulations, you've made it through the ugliness alive!

So, you think you're all set to be your school's resident logician and then, wham, this argument rocks your world:

$$P \to {\sim}Q, P \lor {\sim}R : Q \to (({\sim}P \,\&\, {\sim}R) \lor (Q \to P))$$

Well, at least you know how to start:

1.	$P \to {\sim}Q$	**P**
2.	$P \lor {\sim}R$	**P**

Hmmm, after writing out these premises, there's no doubt about it, this is going to be one hairy proof.

But, what's the harm in giving it the old college try? Here are a few of the statements you come up with along the way:

3.	$Q \to {\sim}P$	1 **Contra**
4.	${\sim}Q \lor {\sim}P$	1 **Impl**
5.	${\sim}(Q \,\&\, P)$	4 **DeM**
6.	${\sim}({\sim}P \,\&\, R)$	2 **DeM**
7.	${\sim}P \to {\sim}R$	2 **Impl**
8.	$R \to P$	7 **Contra**
9.	$R \to {\sim}Q$	1, 8 **HS**
10.	$Q \to R$	9 **Contra**
11.	${\sim}Q \lor R$	10 **Impl**

None of these statements really get you where you need to be. So, it's time for a new tactic — one that's easy to use and that actually works. Yup, you guessed it: The easy-to-use tactic I'm hinting at is conditional proof.

Understanding conditional proof

Conditional proof allows you to use part of the conclusion as a premise that you can use to prove the rest of the conclusion.

To prove the validity of an argument whose conclusion is in the form $x \rightarrow y$ (that is, any $\rightarrow$-statement), you can follow these steps:

1. Detach the sub-statement x.

2. Add x to your list of premises as an *assumed premise* (**AP**).

3. Prove the sub-statement y as if it were the conclusion.

The idea is simple, but brilliant. So, in the example from the previous section, instead of working your way through and trying to prove that the conclusion $Q \rightarrow ((\sim P \& \sim R) \vee (Q \rightarrow P))$ is valid, conditional proof allows you to:

1. Detach the sub-statement Q.

2. Add Q to the list of premises as an **AP**.

3. Prove the sub-statement $(\sim P \& \sim R) \vee (Q \rightarrow P)$ as if it were the conclusion.

The premises are the same, but now you have an additional *assumed premise* (**AP**). Here's how it works:

1.	$P \rightarrow \sim Q$	**P**
2.	$P \vee \sim R$	**P**
3.	Q	**AP**

With this proof, you're trying to build the statement $((\sim P \& \sim R) \vee (Q \rightarrow P))$. This time, however, you have a way to break your premises down and get the pieces you need. For example, check out these steps:

4.	$\sim P$	1, 3 **MT**
5.	$\sim R$	2, 4 **DS**

With all three single-constant statements at your fingertips, you can go to work:

6.	$\sim P \& \sim R$	4, 5 **Conj**
7.	$(\sim P \& \sim R) \vee (Q \rightarrow P)$	6 **Add**

To complete the proof, here's the final formality:

8.	$Q \rightarrow ((\sim P \& \sim R) \vee (Q \rightarrow P))$	3–7 **CP**

This final step is called *discharging the AP*, which makes clear to the reader that, even though you were operating *as if* the assumption were true from statements 3 through 7, you're no longer making this assumption in statement 8. In other words, the conclusion is true even if the assumption *isn't* true!

You may be asking "But I can't do that, can I? I didn't prove that the *actual* conclusion is true. All I proved is that *part* of it is true. And worse yet, I did it using a phony premise that I stole from the conclusion."

You're right. This setup seems too good to be true. After all, premises are like money in the bank, and the conclusion is like a nasty credit card debt you'd rather not look at. But, what if I told you that you could cut your debt in half *and* put money in the bank? That's exactly what conditional proof allows you to do — it's fair, legal, and no credit agency will contact you.

For example, recall that the original argument looked like this:

$$P \rightarrow \sim Q, \sim(\sim P \& R) : Q \rightarrow (\sim(P \vee R) \vee (Q \rightarrow P))$$

If the conclusion could talk, it would say, "You need to show me that if Q is true, then $\sim(P \vee R) \vee (Q \rightarrow P)$ is also true."

Using a conditional proof, you talk back to the conclusion, saying "Okay then, I'll show you that assuming Q is true $\sim(P \vee R) \vee (Q \rightarrow P)$ is also true." And then you do exactly that: You *assume* that Q is true and then you prove the rest.

Tweaking the conclusion

You can apply equivalence rules to the conclusion of an argument to make conditional proof easier to use.

Flipping conclusions with Contra

To take full advantage of this section, you have to first remember that you can use every $\rightarrow$-statement in two ways: The way it is and in its **Contra** form (see Chapter 10). So when the conclusion is a $\rightarrow$-statement, you can use conditional proof to attack it in two different ways.

For example, check out this proof, which is taken at face value:

$$P \rightarrow Q, R \vee (Q \rightarrow P) : \sim(P \leftrightarrow Q) \rightarrow R$$

1.	$P \rightarrow Q$	**P**
2.	$R \vee (Q \rightarrow P)$	**P**
3.	$\sim(P \leftrightarrow Q)$	**AP**

If you begin the proof in this way, your assumed premise is not very helpful.

But, imagine using **Contra** on the conclusion to get $\sim R \to (P \leftrightarrow Q)$. Using **Contra** allows you to take the following steps:

1.	$P \to Q$	**P**
2.	$R \lor (Q \to P)$	**P**
3.	$\sim R$	**AP**

In this case, you're trying to prove $(P \leftrightarrow Q)$. This solution is much more straightforward:

4.	$Q \to P$	2, 3 **DS**
5.	$(P \to Q)\ \&\ (Q \to P)$	1, 5 **Conj**
6.	$P \leftrightarrow Q$	5 **Equiv**

Now, you can discharge your **AP** as follows:

7.	$\sim R \to (P \leftrightarrow Q)$	3–6 **CP**

And don't forget to take the proof all the way to the conclusion:

8.	$\sim(P \leftrightarrow Q) \to R$	7 **Contra**

Any changes you make to the conclusion will appear at the end of the proof, even after you discharge the assumption.

In this case, even though you thought of using **Contra** first, and you wrote the proof with this in mind, it actually appears last.

Winning through implication

You can also turn any ∨-statement into a →-statement by using **Impl** (see Chapter 10). Using **Impl** makes any ∨-statement a potential candidate for conditional proof.

For example, consider this argument:

$P : \sim R \lor (Q \to (P\ \&\ R))$

Without conditional proof, you don't have much hope. But, the problem becomes much simpler after you notice that you can use **Impl** to rewrite the conclusion as

$R \to (Q \to (P\ \&\ R))$

And even better, you can use **Exp** to rewrite it again:

$(R \& Q) \to (P \& R)$

So, now you're ready to go with your proof:

1.	P	**P**
2.	$R \& Q$	**AP**

Now you want to get $P \& R$. These steps practically write themselves:

3.	R	2 **Simp**
4.	$P \& R$	1, 3 **Conj**

With almost no effort, you are ready to discharge your **AP**:

5.	$(R \& Q) \to (P \& R)$	2–4 **CP**

After you discharge the **AP**, the rest is just tracing back to the original form of the conclusion:

6.	$R \to (Q \to (P \& R))$	5 **Exp**
7.	$\sim R \vee (Q \to (P \& R))$	6 **Impl**

Stacking assumptions

After you assume a premise, if the new conclusion is a $\to$-statement (or can be turned into one), you can assume another premise. This is nice way to get two (or more!) assumptions for the price of one. Here's an example:

$\sim Q \vee R : (P \vee R) \to ((Q \& S) \to (R \& S))$

Start out, as usual, with your premise and **AP**:

1.	$\sim Q \vee R$	**P**
2.	$P \vee R$	**AP**

Unfortunately, you still have a long way to go to prove $(Q \& S) \to (R \& S)$. But, because the new conclusion is a $\to$-statement, you can pull out another **AP**, like this:

3.	$Q \& S$	**AP**

Now, the new goal becomes proving *R* & *S*. Here's what the steps look like:

4.	*Q*	3 **Simp**
5.	*S*	3 **Simp**
6.	*R*	1, 4 **DS**
7.	*R* & *S*	5, 6 **Conj**

At this point, you can discharge the last **AP** that you assumed:

8.	(*Q* & *S*) → (*R* & *S*)	3–7 **CP**

And now, you can discharge the first **AP**:

9.	(*P* ∨ *R*) → ((*Q* & *S*) → (*R* & *S*))	2–8 **CP**

When you assume more than one premise, you must discharge them in reverse order: Discharge the *last* premise *first* and then work your way back to the first premise.

Thinking Indirectly: Proving Arguments with Indirect Proof

Just when you thought everything was going your way, you come across an argument that you just can't figure out. For example, consider this argument:

P → (*Q* & ~*R*), *R* : ~(*P* ∨ ~*R*)

1.	*P* → (*Q* & ~*R*)	**P**
2.	*R*	**P**

Seemingly not much you can do with this argument, huh? And because the conclusion is in a form that you cannot turn into a →-statement, conditional proof is out of the picture, so you may think you're really stuck. The good news is that you're not stuck at all. In fact, a whole new world is about to open up to you. And a beautiful world at that!

This section shows you how to use *indirect proof,* which unlike conditional proof, is *always* an option no matter what the conclusion looks like.

Introducing indirect proof

Indirect proof (also called *proof by contradiction*) is a type of logical judo. The idea here is to assume that the conclusion is false and then show why this assumption is wrong. Success means that your conclusion was true all along.

To prove that any argument is valid, you can follow these steps:

1. Negate the conclusion.

2. Add the negation to your list of premises as an assumption.

3. Prove any *contradictory statement* (any statement of the form $x \& \sim x$).

When working on the example from earlier in this section, instead of trying to prove the conclusion, $\sim(P \vee \sim R)$, use indirect proof, which allows you to use its negation, $P \vee \sim R$, as an assumed premise (**AP**). Just remember that now you're looking to prove that this assumption leads to a contradictory statement.

Here's what the proof would look like:

$$P \to (Q \& \sim R), R : \sim(P \vee \sim R)$$

1.	$P \to (Q \& \sim R)$	**P**
2.	R	**P**
3.	$P \vee \sim R$	**AP**

Your goal now is to prove a statement and its negation, then use them to build a contradictory &-statement, like this:

With the **AP** at your disposal, options suddenly open up:

4.	P	2, 3 **DS**
5.	$Q \& \sim R$	1, 4 **MP**
6.	$\sim R$	**Simp**

At this point, you've derived both R and $\sim R$, so you can build them into a single contradictory statement as follows:

7.	$R \& \sim R$	2, 6 **Conj**

The assumption has led to an impossible situation, so you know that the **AP** must be false. If $P \vee \sim R$ is false, then $\sim(P \vee \sim R)$ must be true:

8.	$\sim(P \vee \sim R)$	3–7 **IP**

As with conditional proof, you need to discharge the **AP**, which makes clear that, even though you were operating as if the assumption were true from statements 3 through 7, you're no longer making this assumption in statement 8. In fact, the very thing you've proved is that the assumption *isn't* true!

Proving short conclusions

As I mention in Chapter 9, arguments where the conclusion is shorter than the premises tend to be more difficult to prove than those where the conclusion is longer, because breaking down long premises can be tricky.

However, indirect proof works especially well when the conclusion is shorter than the premises because the negated conclusion becomes a nice short premise for you to use. For example, consider this argument:

$\sim((\sim P \vee Q) \& R) \to S, P \vee \sim R : S$

1.	$\sim((\sim P \vee Q) \& R) \to S$	**P**
2.	$P \vee \sim R$	**P**
3.	$\sim Q \vee S$	**P**

One way or another, you're going to have to break down the first premise, but it will be easier with some help:

4.	$\sim S$	**AP**

Immediately, things open up and you can take the following steps:

5.	$(\sim P \vee Q) \& R$	1, 4 **MT**
6.	$\sim P \vee Q$	5 **Simp**
7.	R	5 **Simp**

Remember, you're trying to derive two contradictory statements. But now, the opportunities are more plentiful:

8.	P	2, 7 **DS**
9.	Q	6, 8 **DS**
10.	S	3, 9 **DS**

When you're doing an indirect proof, don't be fooled into thinking you're done after you prove the conclusion. Remember that you also need to build a contradictory statement.

In this case, the **AP** leads to its own negation, which allows you to complete the proof:

11.	$S \& \sim S$	4, 10 **Conj**
12.	S	4–11 **CP**

Superficially, line 12 looks like line 10, but now you've discharged the **AP**, so the proof is complete.

Combining Conditional and Indirect Proofs

Here's a common question from students: If I'm already using an **AP** for a conditional proof, do I have to start over if I want to add an **AP** for an indirect proof?

The good news is that you don't have to start over, as this example explains:

$$\sim P \& Q \to (\sim R \& S), Q : R \to P$$

1.	$\sim P \& Q \to (\sim R \& S)$	**P**
2.	Q	**P**

On your first pass over the proof, you can only manage to get this:

3.	$\sim (\sim P \& Q) \vee (\sim R \& S)$	1 **Impl**

Maybe the next step is there, but in any case, you're not seeing it. So, you have to move on. And because conditional proof is an option, you try it first:

4.	R	**AP** (for conditional proof)

Now you're trying to prove P, but you're still not sure how to do this directly. Now, you can move on to indirect proof by negating what you're now trying to prove and adding it as a premise:

5.	$\sim P$	**AP** (for indirect proof)

Now the goal is to find a contradiction. Suddenly, the pieces fall into place:

6.	~P & Q	2, 5 **Conj**
7.	~R & S	1, 6 **MP**
8.	~R	7 **Simp**

Now you're ready to discharge the **AP** for the indirect proof:

| 9. | R & ~R | 4, 8 **Conj** |
| 10. | P | 5–9 **IP** |

Of course, proving *P* was the goal of the original conditional proof, so here's what you get:

| 11. | R → P | 4–10 **CP** |

When using conditional and indirect proof methods together, discharge your **AP**s starting with the last one you added and work your way back to the first.

Chapter 12

Putting It All Together: Strategic Moves to Polish Off Any Proof

In This Chapter

▶ Handling easy proofs quickly

▶ Using conditional proof to work through moderate proofs

▶ Giving difficult proofs a run for their money

Some logic problems almost solve themselves. Others look tough at first but soon fall in line when you know what to do. Still others kick and scream every step of the way until you whip them into submission.

This chapter is all about what the Serenity Prayer calls "the wisdom to know the difference." And the wisdom you gain will be the calm confidence you need to write sentential logic (SL) proofs with ease when possible and with perseverance when necessary.

Just as the American justice system declares a defendant "innocent until proven guilty," in this chapter, I recommend the same sort of approach with proofs: "Easy until proven difficult." First, I show you how to make a quick assessment of an argument to get a gut sense of how tough it will be to prove. Next, I describe a few quick-and-dirty moves that are enough to handle the easy proofs in five minutes or less.

For more stubborn proofs, I show you how and when to pull out the conditional proof technique from Chapter 11. The technique is often enough to complete proofs of moderate difficulty.

And finally, for the really tough ones, I show you how to use the form of an SL statement to your advantage by breaking down long premises and working both ends against the middle to complete the proof. I also show you an advanced method of indirect proof.

Easy Proofs: Taking a Gut Approach

You don't need a machine gun to kill a mosquito. Similarly, you don't need to devise a brilliant strategy to solve an easy problem. You just need to spend five minutes looking at what's in front of you and jotting down a few ideas. The following sections show you a few quick tricks for writing simple proofs with grace and speed.

Look at the problem

Looking really does mean just looking. The three suggestions I discuss in the following sections should take less than a minute, but it's a quality minute.

Compare the premises and the conclusion

Do the premises look somewhat similar to the conclusion, or are they very different? If they look similar, the proof may not be so difficult; otherwise, it may be trickier.

In either case, think about what has to happen to bridge the gap. Can you think of any hunches as how to proceed?

That's all this step requires — thinking and working from your gut.

Notice the lengths of the premises and the conclusion

Generally speaking, short premises and a long conclusion indicate an easy proof, whereas long premises and a short conclusion indicate a more difficult proof.

When given enough short premises, you can build almost any conclusion you like. However, on the flip side, breaking down long premises enough to get a short conclusion can be tricky.

In the upcoming section "Jot down the easy stuff," I show you a bunch of ways to break down premises. For now, just get a gut sense of how tough the proof will be based on this rule of thumb: *The shorter the premises, the easier the proof.*

Look for repeated chunks of statements

When you notice chunks of statements that are repeated in an argument, underline them to make them stand out. The best strategy for these chunks is often to leave them alone rather than break them down.

For example, look at this argument:

$(P \leftrightarrow Q) \rightarrow {\sim}(R \& S), R \& S : {\sim}(P \leftrightarrow Q)$

You could hammer away at the statement $(P \leftrightarrow Q) \to \sim(R \& S)$ to break it down, but there's no reason to. After you notice that the chunks $(P \leftrightarrow Q)$ and $(R \& S)$ are repeated in the statement, this one-step solution presents itself:

1. $(P \leftrightarrow Q) \to \sim(R \& S)$ **P**
2. $R \& S$ **P**
3. $\sim(P \leftrightarrow Q)$ 1, 2 **MT**

For now, just noticing these sorts of chunks will keep you speeding on your way.

Jot down the easy stuff

With one minute down, this section gives you four more one-minute strategies to move you along. I call this type of strategy *kitchen-sinking*, because you're using everything in your proof except the kitchen sink.

As you use these strategies, don't hesitate to go for the finish if you see that the way is clear.

Break down statements using Simp and DS

Simp and **DS** are the two simplest rules for breaking down &-statements and ∨-statements (see Chapter 9 for all of your **Simp** and **DS** needs). The more you break down premises early on, the better your chances are for building up the conclusion.

Expand your options using Impl and Contra

Use **Impl** (see Chapter 10) to convert ∨-statements to →-statements, and vice versa. Then use **Contra** to convert every →-statement to its contrapositive. Use these rules to rewrite every statement you can because in either direction, these are easy ways to expand your options for later.

Use MP and MT wherever possible

Opportunities to use **MP** and **MT** (as seen in Chapter 9) are easy to spot and tend to give you simple statements to work with.

Convert all negative statements using DeM

Generally speaking, **DeM** is the only rule that allows you to convert the four negative forms of SL statements to positive ones.

DeM works directly on statements of the forms $\sim(x \& y)$ and $\sim(x \vee y)$, as I cover in Chapter 10. But, even when you're up against →-statements and ↔-statements, the two remaining negative forms, you can use **DeM** after employing a few other rules to get these statements into the forms $\sim(x \& y)$ and $\sim(x \vee y)$.

For example, to convert $\sim(x \to y)$ use these steps:

1.	$\sim(x \to y)$	**P**
2.	$\sim(\sim x \vee y)$	1 **Impl**
3	$x \,\&\, \sim y$	2 **DeM**
4.	x	3 **Simp**
5.	$\sim y$	3 **Simp**

In just a few steps, you've turned a complex-looking statement into two simple ones.

Even when you're saddled with the dreaded form $\sim(x \leftrightarrow y)$, you can use **Equiv** and then break every piece down with **DeM**. For example, take a look at these steps:

1.	$\sim(x \leftrightarrow y)$	**P**
2.	$\sim((x \,\&\, y) \vee (\sim x \,\&\, \sim y))$	1 **Equiv**
3.	$\sim(x \,\&\, y) \,\&\, \sim(\sim x \,\&\, \sim y)$	2 **DeM**
4.	$\sim(x \,\&\, y)$	3 **Simp**
5.	$\sim(\sim x \,\&\, \sim y)$	3 **Simp**

At this point, you can use **DeM** again to simplify statements 4 and 5 still further. Sure, it's a few extra steps, but you've turned an impenetrable statement into two very simple ones.

Know when to move on

Suppose you've spent about five minutes with a particularly hairy problem. You've looked at it and turned it over in your head. You've already jotted down a few simple statements — or maybe there were none to jot down — and now you're just about out of ideas.

My advice: Five minutes. That's all I'd give the gut strategy. And that goes double if the premises are long and the conclusion is short. If the proof doesn't jump out at you in five minutes, you need stronger medicine so that the next five minutes are productive rather than frustrating.

Even if you have to move on to a new tactic, you don't have to start over. Any statement that you've already proved is yours to keep and use throughout the rest of the proof.

Moderate Proofs: Knowing When to Use Conditional Proof

Stronger medicine is coming right up in this section. After you've abandoned hope that the problem is simple, it's time to pull out the conditional proof.

TIP Use conditional proof as your first choice whenever possible because it tends to be the quickest way to a solution for a problem of medium difficulty.

To decide when conditional proof is possible, look at the conclusion you are trying to prove and decide which of the eight basic forms it is. (Check out Chapter 5 for a rundown on all eight basic forms of SL statements.)

You can always use conditional proof for three of the eight basic forms, which I call the *friendly forms*. You can also use conditional proof for two of the forms, which I call the *slightly-less-friendly forms*, but more work is necessary. And finally, you can't use conditional proof for the remaining three forms, which I call the *unfriendly forms*:

Friendly Forms	Slightly-Less-Friendly Forms	Unfriendly Forms
$x \rightarrow y$	$(x \leftrightarrow y)$	$x \& y$
$x \vee y$	$\sim(x \leftrightarrow y)$	$\sim(x \rightarrow y)$
$\sim(x \& y)$		$\sim(x \vee y)$

In this section, I discuss the cases when you can use conditional proof. I save the remaining cases for the section "Difficult Proofs: Knowing What to Do When the Going Gets Tough."

The three friendly forms: $x \rightarrow y$, $x \vee y$, and $\sim(x \& y)$

Obviously, you can always use conditional proof on any conclusion in the form $x \rightarrow y$. But, you can also easily use conditional proof on any conclusion in the forms $x \vee y$ and $\sim(x \& y)$.

REMEMBER You can turn any conclusion of the form $x \vee y$ into a conditional form just by using **Impl**. This rule turns it into $\sim x \rightarrow y$.

For example, suppose you want to prove this argument:

$R : \sim(P \& Q) \vee (Q \& R)$

| 1. | R | P |

Not much is going on here. But, after you realize that the conclusion is equivalent to $(P \& Q) \to (Q \& R)$, you can use conditional proof:

2.	$P \& Q$	AP
3.	Q	2 **Simp**
4.	$Q \& R$	1, 3 **Conj**
5.	$(P \& Q) \to (Q \& R)$	2–4 **CP**

After your AP is discharged, all that's left to do is show how the statement you just built is equivalent to the conclusion you're trying to prove:

| 6. | $\sim(P \& Q) \vee (Q \& R)$ | 5 **Impl** |

The other easy form to work with is $\sim(x \& y)$. When your conclusion is in this form, you need to use **DeM** to get it out of the negative form, changing it to $\sim x \vee \sim y$. From here, use **Impl** to change it to $x \to \sim y$.

For example, suppose you want to prove this argument:

$\sim P \to Q, P \to \sim R : \sim(\sim Q \& R)$

| 1. | $\sim P \to Q$ | P |
| 2. | $P \to \sim R$ | P |

The insight here is to realize that the conclusion is equivalent to $Q \vee \sim R$ (by **DeM**), which is then equivalent to $\sim Q \to \sim R$ (by **Impl**). Again, you can use a conditional proof:

3.	$\sim Q$	AP
4.	P	1, 3 **MT**
5.	$\sim R$	2, 4 **MP**
6.	$\sim Q \to \sim R$	3–5 **CP**

As with previous examples, after you've discharged your AP, you need to complete the bridge from the statement you've built to the conclusion you started with:

| 7. | $Q \vee \sim R$ | 6 **Impl** |
| 8. | $\sim(\sim Q \& R)$ | 7 **DeM** |

The two slightly-less-friendly forms: $x \leftrightarrow y$ and $\sim (x \leftrightarrow y)$

If I've said it once, I've said it a thousand times: Working with $\leftrightarrow$-statements is always a little dicey. But, luckily, the principle here is the same as with the very friendly forms.

Always remember, and please don't ever forget, that the first step with a $\leftrightarrow$-statement is always to get rid of the $\leftrightarrow$-operator by using **Equiv**. Knowing that bit of advice will get you halfway home if you happen to get lost.

To work with a conclusion that's in the form of $x \leftrightarrow y$, you first have to use **Equiv** to change it to $(x \,\&\, y) \lor (\sim x \,\&\, \sim y)$. You should recognize that $\lor$-statement as a very friendly form, which allows you to use **Impl** to change it to $\sim (x \,\&\, y) \to (\sim x \,\&\, \sim y)$.

For example, suppose you want to prove this argument:

$$((\sim P \lor Q) \lor \sim R) \to \sim (P \lor R) : P \leftrightarrow R$$

1.	$((\sim P \lor Q) \lor \sim R) \to \sim (P \lor R)$	**P**

I'm not going to kid you: This is a tough proof. Without finding a way to turn it into a conditional, it's just about hopeless. Fortunately, you can use **Equiv** to change the conclusion to $(P \,\&\, R) \lor (\sim P \,\&\, \sim R)$, and from there, change it to $\sim (P \,\&\, R) \to (\sim P \,\&\, \sim R)$ using **Impl**. Check it out:

2.	$\sim (P \,\&\, R)$	**AP**
3.	$\sim P \lor \sim R$	2 **DeM**

Now, you're trying to prove $\sim P \,\&\, \sim R$. The big question at this point is "How do I use that hairy premise in line 1?" (This is why I say that long premises make for difficult proofs.) Okay, first things first: You can at least unpack the premise a little bit using **DeM** on the second part:

4.	$((\sim P \lor Q) \lor \sim R) \to (\sim P \,\&\, \sim R)$	1 **DeM**

The second part of this statement looks just like what you're trying to prove for your conditional proof. So, if you can build the first part of this statement, you can use **MP** to get the second part. The goal now, then, is to prove the statement $(\sim P \lor Q) \lor \sim R$.

But, now you're probably wondering how you get a Q from nothing at all. The great insight here is that you can use **Add** to tack a Q onto $\sim P \lor \sim R$:

5.	$(\sim P \lor \sim R) \lor Q$	3 **Add**

The next part of this proof requires just a bunch of manipulation using **Assoc** and **Comm**:

6.	$\sim P \vee (\sim R \vee Q)$	5 **Assoc**
7.	$\sim P \vee (Q \vee \sim R)$	6 **Comm**
8.	$(\sim P \vee Q) \vee \sim R$	7 **Assoc**

Finally, you get a glimpse of the Holy Grail: You can use line 8 together with line 4 to derive what you need to discharge your premise:

9.	$\sim P \,\&\, \sim R$	4, 8 **MP**
10.	$\sim(P \,\&\, R) \to (\sim P \,\&\, \sim R)$	2–9 **CP**

As usual, after discharging the **AP**, you need to make the statement that you've just built look like the conclusion:

11.	$(P \,\&\, R) \vee (\sim P \,\&\, \sim R)$	10 **Impl**
12.	$P \leftrightarrow R$	11 **Equiv**

Okay, so that was a bear of a proof. But turning that tough conclusion into a more manageable form made it totally possible.

Proving conclusions that are in these slightly-less-friendly forms can get tough. This example is certainly pushing the envelope of what I'd call medium difficulty. I give an example of how to prove a conclusion in the form $\sim(x \leftrightarrow y)$ later in this chapter, in the "Difficult Proofs: Knowing What to Do When the Going Gets Tough" section.

The three unfriendly forms: $x \,\&\, y$, $\sim(x \vee y)$, and $\sim(x \to y)$

When the conclusion falls into this category, conditional proof is almost never an option because, as a rule, you can't turn these three forms into the form $x \to y$.

To understand why you can't use conditional proof in these cases, first notice that no rule exists for turning a statement that's in the form $x \,\&\, y$ into a $\to$-statement. Similarly, statements in the two remaining unfriendly forms are easy to turn into &-statements, but again you get stuck if you try to turn them into $\to$-statements.

Whenever you're faced with a conclusion that's in an unfriendly form, my advice is to move on and attempt a direct or indirect proof. In either case, you're probably looking at a difficult proof.

Difficult Proofs: Knowing What to Do When the Going Gets Tough

When a gut attempt at a direct proof fails, and you can't use conditional proof, you're probably facing a difficult proof. In these cases, you're faced with a choice of trying to push forward with a direct proof or switching over to an indirect proof.

The upcoming sections show you some strategies that help push you through the really tough proofs.

Choose carefully between direct and indirect proof

If you come across a problem whose conclusion can't be turned into a → statement (which would allow you to use conditional proof), consider direct proof your first choice *except* when you hit one of the three exceptions that I list in this section.

Even if you don't find a direct proof in the first five minutes, stick with it for a while longer when you can't use a conditional proof. In the best case scenario, using some of the ideas later in this chapter, you may find a direct proof without switching to indirect proof. And at worst, if you need to switch to indirect proof, you can still use all of the statements you came up with while looking for a direct proof.

This switcheroo is not true in the reverse direction: If you switch to indirect proof too early and then later want to abandon ship and try direct proof again, you can't use any of the statements you found while looking for an indirect proof.

Exception #1: When the conclusion is short

Indirect proof can be very helpful when the conclusion is a short statement and most or all of the premises are long. In this case, turning a troublesome short conclusion into a helpful short premise is a win-win situation.

Exception #2: When the conclusion is long but negated

Indirect proof is especially good for handling conclusions that are in negative forms — $\sim(x \lor y)$ and $\sim(x \to y)$ — because a negated conclusion will become a positive premise. For example:

$P, Q : \sim((Q \lor R) \to (\sim P \& R))$

1.	P	**P**
2.	Q	**P**
3.	$(Q \lor R) \to (\sim P \& R)$	**AP**

Using indirect proof changes the conclusion to a positive form and then adds it as a premise. Now you can begin to chip away at it:

4.	$Q \lor R$	2 **Add**
5.	$\sim P \& R$	3, 4 **MP**
6.	$\sim P$	5 **Simp**
7.	$P \& \sim P$	1, 6 **Conj**
8.	$\sim((Q \lor R) \to (\sim P \& R))$	3–8 **IP**

Exception #3: When all hope seems lost

The third place to use an indirect proof is when you've banged away at it with direct proof for a while and you're not getting anywhere. Just wait a while is all I ask. Converting from direct to indirect proof is always easy. And even if you're using a conditional proof already, you can always convert to indirect proof just by adding on another **AP**. (See Chapter 11 for an example of how **AP** works.)

Work backwards from the conclusion

I've compared writing a proof to building a bridge to get you from here to there. And usually, that's exactly what happens: You're able to get from here to there. But, sometimes, it doesn't work out as well. So, if you find yourself in a bind and you can't get there from here, it may be easier to get *here* from *there*.

In Chapter 9, I show you a quick example of working backwards from the conclusion. In this section, you use this skill to write a very difficult proof. For example, suppose you want to prove this argument:

$P \to Q, (P \to R) \to S, (\sim Q \lor \sim S) \& (R \lor \sim S) : \sim(Q \leftrightarrow R)$

The conclusion is of the form $\sim(x \leftrightarrow y)$, one of the two "slightly-less-friendly" forms. So if you want to use a conditional proof, you're going to have to rearrange the conclusion. Here, writing down the *end* of the proof before the beginning is a good idea:

99.   ~$(Q \leftrightarrow R)$

You begin by numbering the last line of the proof *99*. You're not going to need that many lines, but you can renumber them at the end. As you can see, the last line contains the conclusion in all its glory.

Then start thinking backwards from the conclusion about how you may have ended up at that conclusion in particular. In this case, the best way to do the proof is to use conditional proof, which means that the conclusion has to be a →-statement. And the first step was to use **Equiv** to turn the statement into the friendly form ~$(x \& y)$:

98.   ~$((Q \rightarrow R) \& (R \rightarrow Q))$

99.   ~$(Q \leftrightarrow R)$ 98 **Equiv**

Now, you have to decide what step came just before the one you just figured out. This time, you use **DeM** to get statement 97 out of its negative form:

97.   ~$(Q \rightarrow R) \vee$ ~$(R \rightarrow Q)$

98.   ~$((Q \rightarrow R) \& (R \rightarrow Q))$ 97 **DeM**

99.   ~$(Q \leftrightarrow R)$ 98 **Equiv**

After you get the hang of it, you can see that the next step is to turn statement 97 into a →-statement by using **Impl**:

95.   ~$(R \rightarrow Q)$

96. $(Q \rightarrow R) \rightarrow$ ~$(R \rightarrow Q)$ 4–95 **CP**

97.   ~$(Q \rightarrow R) \vee$ ~$(R \rightarrow Q)$ 96 **Impl**

98.   ~$((Q \rightarrow R) \& (R \rightarrow Q))$ 97 **DeM**

99.   ~$(Q \leftrightarrow R)$ 98 **Equiv**

After all that, you now have the Holy Grail in your hot little hands. You know how the story ends: Having assumed $Q \rightarrow R$ and used it to build ~$(R \rightarrow Q)$, you discharge this assumption by joining these two sub-statements using the →-operator.

And now, of course, you know a lot more about how the story begins. In particular, you know that your **AP** should be $Q \rightarrow R$. So, you know your beginning steps would look like this:

1. $P \rightarrow Q$ **P**

2. $(P \rightarrow R) \rightarrow S$ **P**

3. $($~$Q \vee$ ~$S) \& (R \vee$ ~$S)$ **P**

4. $Q \rightarrow R$ **AP**

Now the goal is building the statement $\sim(R \to Q)$, which is the last backward step you figured out. When you look at your AP and line 1, something probably jumps out at you:

5.	$P \to R$	1, 4 **HS**

Then you look at line 2 and you feel like it's your birthday:

6.	S	2, 5 **MP**

What now? Line 3 is the only premise left, and the form probably looks very, very familiar:

7.	$(\sim Q \,\&\, R) \vee \sim S$	3 **Dist**

Then you get another break:

8.	$\sim Q \,\&\, R$	6, 7 **DS**

When you hit this point in the proof, it pays to know how to manipulate the eight basic forms:

9.	$\sim(Q \vee \sim R)$	8 **DeM**
10.	$\sim(\sim R \vee Q)$	9 **Comm**
11.	$\sim(R \to Q)$	10 **Impl**

At this point, you've reached your goal, and all you need to do is renumber the last few rows:

12.	$(Q \to R) \to \sim(R \to Q)$	4–11 **CP**
13.	$\sim(Q \to R) \vee \sim(R \to Q)$	12 **Impl**
14.	$\sim((Q \to R) \,\&\, (R \to Q))$	13 **DeM**
15.	$\sim(Q \leftrightarrow R)$	14 **Equiv**

Go deeper into SL statements

By now, you're probably getting good at noticing which of the eight forms a statement falls into. Sometimes, though, this isn't quite enough. When a proof is difficult, it often depends on your understanding the structure of statements at a deeper level.

You may have noticed that three of the equivalence rules break a statement into three parts (x, y, and z) rather than just two. These rules are **Exp**, **Assoc**, and **Dist** (see Chapter 10). You probably haven't used these as much as some of the other rules, but for the tough proofs you definitely need them. After you start looking for opportunities to use these three rules, you'll find them all around you.

For example, look at these three statements:

$(P \lor Q) \lor R$

$(P \& Q) \lor R$

$(P \& Q) \lor (R \lor S)$

Can you spot that all of these statements both have the same basic form $x \lor y$? But, don't let this similarity in basic structure blind you to important differences in the deeper structure of these statements.

For example, you can apply **Assoc** to the first statement but not the second, and **Dist** to the second but not the first. And you can apply both **Assoc** and **Dist** to the third statement. These differences become essential when proofs become difficult. Read this section to get some tips on noticing and exploiting these differences.

Using Exp

Exp: $x \rightarrow (y \rightarrow z)$ is equivalent to $(x \& y) \rightarrow z$

For example, check out this statement:

$(P \& Q) \rightarrow (R \rightarrow S)$

This statement is in the form $x \rightarrow y$. But you can also look at it in two other ways. One possibility, taking $(P \& Q)$ as a single chunk, is to notice that the statement looks like:

$x \rightarrow (y \rightarrow z)$

In this case, you can use **Exp** to change the statement to

$((P \& Q) \& R) \rightarrow S$

A second option, taking $(R \rightarrow S)$ as a single chunk, is to notice that the statement looks like

$(x \& y) \rightarrow z$

This time, you can use **Exp** in the other direction, like this:

$P \rightarrow (Q \rightarrow (R \rightarrow S))$

Combining Assoc with Comm

Assoc: $(x \& y) \& z$ is equivalent to $x \& (y \& z)$

$(x \lor y) \lor z$ is equivalent to $x \lor (y \lor z)$

For example, consider this statement:

$\sim(P \lor Q) \rightarrow (R \lor S)$

One way to go here is to apply **Impl** so that you get this:

$(P \lor Q) \lor (R \lor S)$

Now, you can apply **Assoc** in two different directions:

$P \lor (Q \lor (R \lor S))$

$((P \lor Q) \lor R) \lor S$

You can also use **Comm** to rearrange the variables in a bunch of different ways. For example, just working with $P \lor (Q \lor (R \lor S))$, you can get:

$P \lor (Q \lor (S \lor R))$

$P \lor ((R \lor S) \lor Q)$

$(Q \lor (R \lor S)) \lor P$

If you can express a statement using only $\lor$-operators (or only &-operators), you can use a combination of **Assoc** and **Comm** to rearrange the variables in any order you like, which can be a very powerful tool that helps you shape a statement into just about anything you need.

Getting Dist

Dist: $x \& (y \lor z)$ is equivalent to $(x \& y) \lor (x \& z)$

$x \lor (y \& z)$ is equivalent to $(x \lor y) \& (x \lor z)$

Dist also has two other forms worth knowing, with the sub-statement in parentheses at the front:

$(x \lor y) \& z$ is equivalent to $(x \& z) \lor (y \& z)$

$(x \& y) \lor z$ is equivalent to $(x \lor z) \& (y \lor z)$

Most professors find these other two forms of the rule acceptable. However, a few sticklers may make you use **Comm** to turn $(x \lor y) \& z$ into $z \& (x \lor y)$ before applying **Dist**. Or, similarly, they may require you to use **Comm** to turn $(x \& y) \lor z$ into $z \& (x \& y)$ before applying **Dist**.

For example, suppose you have this statement:

$P \lor (Q \mathbin{\&} (R \lor S))$

You can use **Dist** two ways here. First, taking $(R \lor S)$ as a single chunk, notice that the statement looks like this:

$x \lor (y \mathbin{\&} z)$

Now you can rewrite the statement so that it looks like this:

$(P \lor Q) \mathbin{\&} (P \lor (R \lor S))$

The advantage here is that you can use **Simp** to separate the statement into two smaller statements:

$P \lor Q$

$P \lor (R \lor S)$

The second option is to use **Dist** on the sub-statement $Q \mathbin{\&} (R \lor S)$, which would give you

$P \lor ((Q \mathbin{\&} R) \lor (Q \mathbin{\&} S))$

Now, you have three sub-statements — P, $(Q \mathbin{\&} R)$, and $(Q \mathbin{\&} S)$ — joined by $\lor$-statements, which means that you can use **Assoc** and **Comm** to arrange them in any order you like (as long as you keep the stuff inside the parentheses intact).

One great use of **Dist** is that it allows you to change the main operator of a statement from & to $\lor$. By changing the main operator in this way, you can change an unfriendly conclusion — that is, one on which you can't use a conditional proof — to a friendly one.

For example, consider this statement:

$P \mathbin{\&} (Q \lor R)$

In most cases, when you're faced with a conclusion in the form $x \mathbin{\&} y$, you can forget about conditional proof. But, in this case, you can use **Dist** to change it to:

$(P \mathbin{\&} Q) \lor (P \mathbin{\&} R)$

Now, you have the conclusion in a friendly form, and you can use **Impl** to turn it into a $\rightarrow$-statement:

$\sim(P \mathbin{\&} Q) \rightarrow (P \mathbin{\&} R)$

Similarly, another unfriendly form is $\sim(x \rightarrow y)$. Knowing this form may cause you to abandon conditional proof on the following conclusion:

$\sim(P \rightarrow \sim(Q \lor R))$

But, fortunately, you can get this conclusion out of its negative form in two steps. First, use **Impl** to make it look like this:

$$\sim(\sim P \vee \sim(Q \vee R))$$

Next, use **DeM**:

$$P \,\&\, (Q \vee R)$$

Surprisingly, this conclusion is the same as the one I started out with in the $P \,\&\, (Q \vee R)$ example (five statements above this one), so you can use **Dist** as you did there to transform it to a friendly form, and then attack it with a conditional proof.

Break down long premises

As I've mentioned several times, breaking down long premises can be difficult. Sometimes, though, there's no way around it. For example, take a look at this argument:

$$(P \,\&\, Q) \vee (Q \rightarrow R),\ Q,\ \sim R : P$$

1.	$(P \,\&\, Q) \vee (Q \rightarrow R)$	**P**
2.	Q	**P**
3.	$\sim R$	**P**

The key to this one — whether you go with a direct or an indirect proof — is to find a way to break down that long premise. I suggest a direct proof.

When working with a long premise, decide which form of SL statement it is. Doing so often helps suggest the next step you should take.

The form of the first premise is $x \vee y$. When faced with a $\vee$-statement, first try using **Impl**:

4.	$\sim(P \,\&\, Q) \rightarrow (Q \rightarrow R)$	1 **Impl**

Then you can look at this statement as $x \rightarrow (y \rightarrow z)$, which allows you to try **Exp**:

5.	$(\sim(P \,\&\, Q) \,\&\, Q) \rightarrow R$	4 **Exp**

You're in good shape now because you've isolated R as the second part of the statement. So, you can use **MT**:

6.	$\sim(\sim(P \,\&\, Q) \,\&\, Q)$	3, 5 **MT**

Be careful that you don't try to cancel out the two ~-operators here. The first one applies the entire statement, whereas the second one applies only to the sub-statement (*P* & *Q*).

Now, the form of the statement is ~(*x* & *y*), which means you've come across a good time to use **DeM**:

7.	(*P* & *Q*) ∨ ~*Q*	6 **DeM**

The rest of the steps suggest themselves:

8.	*P* & *Q*	2, 7 **DS**
9.	*P*	8 **Simp**

Note that keeping track of the form of the statement was instrumental in peeling away layer after layer of this premise. Sometimes, when students see tough proofs like these, they ask, "But what if I just don't see it? And what if I can't figure out the next step?"

The good news is that you can almost always find more than one way to do a proof. So, I'm going to show you that even if you start out differently, you can usually still find a way to make it work.

Here's the same proof that I just discussed, but with a different opening step:

1.	(*P* & *Q*) ∨ (*Q* → *R*)	**P**
2.	*Q*	**P**
3.	~*R*	**P**

In this case, start by applying **Impl** not to the main operator but to the second part of the statement:

4.	(*P* & *Q*) ∨ (~*Q* ∨ *R*)	1 **Impl**

This time, you can look at this statement as a form of *x* ∨ (*y* ∨ *z*), which means you can use **Assoc**:

5.	((*P* & *Q*) ∨ ~*Q*) ∨ *R*	4 **Assoc**

Surprise, surprise — you're again in the position to use **DS**. However, this time you can use it twice in a row:

6.	(*P* & *Q*) ∨ ~*Q*	3, 5 **DS**
7.	*P* & *Q*	2, 6 **DS**

And, once again, the answer emerges:

8.	*P*	7 **Simp**

Make a shrewd assumption

In Chapter 11, I show you how to use indirect proof by assuming the negation of the conclusion and then disproving it.

But, with indirect proof, you aren't limited to using the negation of the conclusion. In fact, you can make *any* assumption and try to disprove it. If you are successful, then you have proved the *negation* of the assumption, and this can often help you prove the conclusion.

Although you can make any assumption, the strategy here is to pick an assumption that will quickly lead to a contradiction. For example, here is the argument that I proved valid in the earlier section "Break down long premises":

$(P \& Q) \vee (Q \to R), Q, \sim R : P$

1.	$(P \& Q) \vee (Q \to R)$	P
2.	Q	P
3.	$\sim R$	P

In this proof, you're looking for a quick way to break down the first premise. This time, you create an assumed premise out of thin air that will help make this happen:

| 4. | $\sim(P \& Q)$ | AP |

As with all indirect proofs, now you're looking for a contradiction. But, this takes only a few lines:

5.	$Q \to R$	1, 4 **DS**
6.	R	2, 5 **MP**
7.	$R \& \sim R$	3, 6 **Conj**

As usual, the next step is to discharge your **AP**:

| 8. | $P \& Q$ | 4-7 **IP** |

Now, completing the proof is almost trivial:

| 9. | P | 8 **Simp** |

Chapter 13

One for All and All for One

*I*n Chapter 4, I show you that the word *or* has two different meanings in English: The inclusive *or* means "either this or that, or both," whereas the exclusive *or* means "either this or that, but not both."

I also note that the ∨-operator in sentential logic (SL) removes this ambiguity because it always represents the inclusive *or*. At the time, you may have thought that this was highly unfair and discriminatory. In fact, the more rebellious among you may have thought about starting a movement to add a sixth operator to SL.

But, before you take to the streets carrying homemade signs and chanting "Two, four, six, eight, the exclusive *or* is really great," read this chapter. In these pages, you find out how the exclusive *or* — as well as any other home-made operator you may come up with — is already covered by the five SL operators. In fact, in this chapter, I show you how these five symbols allow you to express *any* possible truth function you care to devise.

You also discover how you can express any possible truth function with *fewer* than five SL operators. In fact, you may be surprised how much you can do with so little.

Making Do with the Five SL Operators

After you've been working with the five SL operators for a while, you may begin to wonder whether a few more operators would make the language more useful. In this section, I show you that the answer is a resounding "No!"

In order to demonstrate this point, I invent a new fictional operator — the ?-operator — for the sake of this discussion. The ?-operator will work just like an exclusive *or* (see Chapter 4), which means *either . . . or . . . but not both.* Here's a truth table for this new operator:

x	y	x ? y
T	T	F
T	F	T
F	T	T
F	F	F

You could then use this newfangled operator just like you would use any other operators in SL. For example, you could combine it with the old familiar operators to make a statement like $(P\,?\,Q) \to P$.

You could even use a truth table to discover under which interpretations the value of this statement is **T** or **F**:

P	Q	(P	?	Q)	→	P
T	T	T	F	T	**T**	T
T	F	T	T	F	**T**	T
F	T	F	T	T	**F**	F
F	F	F	F	F	**T**	F

It seems like a great idea. So why hasn't it caught on? Well, because you don't need it. You can get the same result using only standard SL operators:

x	y	x?y	~	(x	↔	y)
T	T	**F**	**F**	T	T	T
T	F	**T**	**T**	T	F	F
F	T	**T**	**T**	F	F	T
F	F	**F**	**F**	F	T	F

As you can see, the statements $x \; ? \; y$ and $\sim(x \leftrightarrow y)$ are semantically equivalent (see Chapter 6 for more on semantic equivalence). And when two statements are equivalent, you can substitute one for the other.

For example, you can substitute the statement $\sim(P \leftrightarrow Q) \to P$ for the statement $(P \; ? \; Q) \to P$. As the following table verifies, these two statements are also semantically equivalent:

P	Q	(P	?	Q)	→	P	~	(P	↔	Q)	→	P
T	T	T	F	T	**T**	T	F	T	T	T	**T**	T
T	F	T	T	F	**T**	T	T	T	F	F	**T**	T
F	T	F	T	T	**F**	F	T	F	F	T	**F**	F
F	F	F	F	F	**T**	F	F	F	T	F	**T**	F

So, you really don't need a ?-operator to represent the exclusive *or*. The five operators are sufficient for all your exclusive-or needs.

In fact, *any* operator you could possibly invent would be equally unnecessary. That is, the five SL operators are sufficient to represent any statement you may wish to express in SL without resorting to additional operators.

Downsizing — A True Story

In the preceding section, I showed you that the ?-operator was unnecessary by showing you how to represent the statement $(x ? y)$ as $\sim(x \leftrightarrow y)$. In other words, I showed you that these two statements are semantically equivalent.

When two statements are semantically equivalent, you can substitute one for another whenever you choose. This swapping power comes in handy when you're doing proofs. It also causes an unexpected result, as the fable in the upcoming sections illustrates.

In this section, you'll use your knowledge of the equivalence rules (see Chapter 10) to see how you can eliminate SL operators and still express what you need to express in SL.

The tyranny of power

Suppose you're just getting tired of the $\leftrightarrow$-operator. It's late to work all the time and is always taking sick days. You want to pull a Donald Trump and say "You're fired!" but you're worried that the other four operators won't be able to do the job by themselves. Then you get an idea.

Using the **Equiv** rule (see Chapter 10), any statement of the form $x \leftrightarrow y$ is equivalent to $(x \,\&\, y) \vee (\sim x \,\&\, \sim y)$. Because of this equivalence, you decide to give the other operators fancy new titles and shift around their responsibilities. From now on, instead of using the $\leftrightarrow$-operator, you'll use a combination of other variables and operators.

For example, instead of writing

$P \,\&\, (Q \leftrightarrow R)$

you'll write

$P \,\&\, ((Q \,\&\, R) \vee (\sim Q \vee \sim R))$.

Similarly, instead of writing

$(P \rightarrow \sim Q) \leftrightarrow (R \,\&\, S)$

you'll write

$((P \rightarrow \sim Q) \,\&\, (R \,\&\, S)) \vee (\sim(P \rightarrow \sim Q) \,\&\, \sim(R \,\&\, S))$

It's a little awkward, but it works. In fact, the great discovery here is that you're now able to do everything with four operators that you used to do with five. And when I say everything, I mean *everything*.

The blow of insurrection

It's lonely at the top — even for logic operators. First thing Monday morning, the →-operator shuffles into your office without knocking. It doesn't like the new arrangement one bit, and after much shouting, it gives you an ultimatum: "Either rehire the ↔-operator or fire me!"

Of course, you don't take kindly to that type of talk, so you have the →-operator escorted out of the building by a burly pair of parentheses. (This is officially where the story takes on an allegorical life all its own.) After the heat of the moment cools a bit, you realize you have another hole to fill.

But again, with the **Impl** rule (see Chapter 10), you can always replace any statement of the form $x \rightarrow y$ with the equivalent statement $\sim x \vee y$. This means, for example, that you can rewrite the statement

$$(((P \rightarrow Q) \,\&\, R) \rightarrow S)$$

as

$$(\sim((\sim P \vee Q) \,\&\, R) \vee S)$$

And again, the whole system is up and running. You can do everything with three operators ($\sim$, $\&$, and $\vee$) that you used to do with five.

These three operators are sufficient to express all SL statements, which is of major importance in Boolean algebra, the first rigorous formalization of logic. You can read more about Boolean algebra in Chapter 14.

The horns of dilemma

Just when things seem back to normal in the operator fable, the &-operator and the ∨-operator request a meeting. They're working longer hours and they know you need them, so they both want big raises.

You offer the operators cigars and brandy and tell them you'll place their request at the top of the agenda at the next Board of Directors' meeting. After they're out of your office, you devise a plan to get rid of one of them so that you can give the other a 50 percent raise and still come out ahead on the deal.

Using DeMorgan's Laws (**DeM** from Chapter 10), you realize that you can replace any statement of the form $x \& y$ with the equivalent statement $\sim(\sim x \lor \sim y)$. But, then again, you can also replace any statement of the form $x \lor y$ with the equivalent statement $\sim(\sim x \& \sim y)$.

For the first time, you hesitate. You even consider firing both of them and rehiring the $\rightarrow$-operator (figuring that you could replace $x \& y$ with $\sim(x \rightarrow \sim y)$ and replace $x \lor y$ with $\sim x \lor y$).

In any case, you now have a lot of room to negotiate. The following three combinations would allow just two operators to do the job of all five:

$\sim$ and $\&$

$\sim$ and $\lor$

$\sim$ and $\rightarrow$

The (Sheffer's) stroke of genius

You didn't see it coming. In walks your most faithful employee, the $\sim$-operator, to give one month's notice. And this time, it isn't about the money, the hours, or even the office mood. It wants nothing more, in fact, than an early retirement with a halfway decent settlement package.

With this news, you may have to close up shop for good. Even if all four operators came back, you can't negate a statement without the $\sim$-operator.

Just when everything looks most unfortunate, a surprise visitor appears: the $|$-operator. The $|$-operator is pronounced *nand operator* (short for *not and*). It's sometimes called *Sheffer's stroke*, after inventor, Henry Sheffer. The statement $x \mid y$ is semantically equivalent to the statement $\sim(x \& y)$.

x	y	$x\mid y$	$\sim$	$(x$	$\&$	$y)$
T	T	**F**	**F**	T	T	T
T	F	**T**	**T**	T	F	F
F	T	**T**	**T**	F	F	T
F	F	**T**	**T**	F	F	F

Hiring on this new help provides some distinct advantages. For example, with this operator, you can express a negative expression, such as ~x, using the expression $x \mid x$:

x	~x	$x \mid x$
T	F	F
F	T	T

You can also express &-statements using the expression $(x \mid y) \mid (x \mid y)$. Additionally, you can express ∨-statements, such as $x \vee y$, using the expression $(x \mid x) \mid (y \mid y)$.

In fact, Sheffer's stroke allows you to express all that the five SL operators allow by using just this one operator. For example, take the expression

$$X \rightarrow (Y \leftrightarrow R)$$

Start out by dropping the ↔-operator:

$$X \rightarrow ((Y \& R) \vee (\sim Y \vee \sim R))$$

Then take care of the → operator:

$$\sim X \vee ((Y \& R) \vee (\sim Y \vee \sim R))$$

Now take care of the &-operators and ∨-operators:

$$\sim X \vee (((Y \mid R) \mid (Y \mid R)) \vee (\sim Y \vee \sim R))$$

$$\sim X \vee (((Y \mid R) \mid (Y \mid R)) \vee (\sim Y \mid \sim Y) \mid (\sim R \mid \sim R))$$

$$\sim X \vee ((((Y \mid R) \mid (Y \mid R)) \mid ((Y \mid R) \mid (Y \mid R))) \mid (((\sim Y \mid \sim Y) \mid (\sim R \mid \sim R)) \mid ((\sim Y \mid \sim Y) \mid (\sim R \mid \sim R))))$$

$$(\sim X \mid \sim X) \mid (((((Y \mid R) \mid (Y \mid R)) \mid ((Y \mid R) \mid (Y \mid R))) \mid (((\sim Y \mid \sim Y) \mid (\sim R \mid \sim R)) \mid ((\sim Y \mid \sim Y) \mid (\sim R \mid \sim R)))) \mid ((((Y \mid R) \mid (Y \mid R)) \mid ((Y \mid R) \mid (Y \mid R))) \mid (((\sim Y \mid \sim Y) \mid (\sim R \mid \sim R)) \mid ((\sim Y \mid \sim Y) \mid (\sim R \mid \sim R)))))$$

Finally, handle the ~-operators:

$$((X \mid X) \mid (X \mid X)) \mid (((((Y \mid R) \mid (Y \mid R)) \mid ((Y \mid R) \mid (Y \mid R))) \mid ((((Y \mid Y) \mid (Y \mid Y)) \mid ((R \mid R) \mid (R \mid R))) \mid (((Y \mid Y) \mid (Y \mid Y)) \mid ((R \mid R) \mid (R \mid R))))) \mid (((((Y \mid R) \mid (Y \mid R)) \mid ((Y \mid R) \mid (Y \mid R))) \mid ((((Y \mid Y) \mid (Y \mid Y)) \mid ((R \mid R) \mid (R \mid R))) \mid (((Y \mid Y) \mid (Y \mid Y)) \mid ((R \mid R) \mid (R \mid R))))))$$

All right, so it's a little tedious, but it can be done. So, after much thought, you proclaim "You're hired!" Cue the flashing lights and applause.

The moral of the story

Of course, in reality the five SL operators are in little danger of being fired, and I highly doubt they'll be retiring anytime soon, which is a darn good thing. Even though the |-operator is logically capable of handling all of SL, the final example in the previous section shows you just how eyeball-bending these statements would be without the other operators.

To make an analogy, if you're computer savvy, you know that everything you do on a computer is translated into 1s and 0s. Your computer keypad, however, contains far more than just these two keys.

This redundancy may be logically unnecessary, but it makes using a computer much easier (just like the five SL operators make logic so much easier). You can use your natural language rather than the computer's language.

And, in a similar way, the five SL operators closely parallel words such as *and, or, not,* and so on, to make it easier for you to think about the meaning of what you're doing rather than the rules for doing it.

Chapter 14

Syntactical Maneuvers and Semantic Considerations

S teve Martin once remarked that you never hear people say:

"Hand me that piano."

The reason this statement is funny is because on one level it's a perfectly normal sentence and on another level it's completely absurd. This chapter is all about these two levels, which include the following:

✔ **Syntax:** *Syntax,* which is the level of grammar, is where Steve Martin's sentence masquerades as normal. After all, "Hand me that piano." is a grammatical sentence, from the capital letter at the beginning to the period at the end. Syntax is all about the form, or internal structure, of language. It's all about the rules that make language possible.

✔ **Semantics:** *Semantics,* which is the level of meaning, is where Steve Martin's sentence reveals itself to be ridiculous. In other words, you can hand someone a pen, a wallet, or even a skunk, but you just can't hand a person a piano. Semantics is all about the function, or external usage, of language. Here, it's all about the meaning that makes language worthwhile.

Whether you're describing a natural language, such as English, or an invented language, such as sentential logic (SL), syntax and semantics both play vital roles. In this chapter, I clarify this key distinction between syntax and semantics in SL. I discuss the rules for well-formed formulas in SL, which describe how to tell an SL statement from a mere string of symbols. Finally, I introduce Boolean algebra, which is an earlier system of logic.

Are You WFF Us or Against Us?

I have an easy question for you. Which of the following is a statement in SL?

A) $(P \lor Q) \to \sim R$

B) $\lor R \, Q \to) \sim (P$

If you chose statement *A*, you're correct. Now, a more difficult question: How did you know?

You may be tempted to respond with a resounding "Uh, duh!" However, it's worth noticing that *A* and *B* contain all the same symbols. They even contain the same number of spaces between the symbols (four, but who's counting?) Okay, so *B* looks like *A* after it went through the blender. But, is that really a good enough reason to reject it as an SL statement? Actually, it's a very good reason.

SL, like any other language, is more than just a bunch of symbols thrown together in any order. The order of its symbols is just one of the things that allow a language to function. And if order weren't so important, you could write an English sentence like this: Dog food than Empire State Building cobra wine goblet Barbra Streisand glass pagoda fdao udos keowe !voapa-aifaoidao-faid, ; s; j?jj;ag u,R.

Yeah, you're right. That last "sentence" makes no sense. It should remind you, however, that the English language also has rules for how you can combine its symbols. These rules are commonly called grammar, and even though you may say that you don't have the faintest clue about it, even as you read this book, your brain is crunching through these rules with great speed and accuracy.

Grammar — also known as syntax — is simply a set of rules for how discrete chunks of language can be ordered. In written language, these chunks are called words and punctuation. In SL, these chunks are called constants, operators, and parentheses.

In this section, you'll learn how to tell a statement in SL from a clever pretender posing as a statement. You'll learn the precise rules for building a string of SL symbols to find out whether it's a statement.

Understanding WFFs (with a few strings attached)

Any random series of SL symbols is called a *string*. Some strings are statements and others aren't. Deciding whether a string is a statement is a *syntactic* question — that is, a question about the form of the string.

When you're talking about syntax, replacing the word *statement* with the term *well-formed formula* is common. Notice that the word *form* appears twice here to hammer home the understanding that the form (in other words, the syntax) of the string is in question. The phrase *well-formed formula* is shortened to WFF, which gets pronounced "wiff" by logicians in the know.

Every string of SL symbols is either a WFF or a non-WFF.

Make no mistake: In SL, the words *statement* and *WFF* mean exactly the same thing. But, when you get to talking about strings at your next cocktail party or logician soiree, you'll want to say *WFF* instead of *statement*, so that's what I do here.

You already know intuitively that $(P \lor Q) \to \sim R$ is a WFF but that $\lor RQ \to) \sim (P$ isn't. And, in this case, your intuition is correct — but what exactly fuels this intuition? Here, I show you three simple rules for turning that intuition into knowledge.

With these three simple rules, you can build any of the SL statements you have seen in this book, and any you will ever see in the future. And just as important, you're prevented from building strings that aren't WFFs. Without further ado, here are the three life-saving rules:

- **Rule 1:** Any constant (*P, Q, R*, and so on) is a WFF.
- **Rule 2:** If any string *x* is a WFF, then the string $\sim x$ is also a WFF.
- **Rule 3:** If any two strings *x* and *y* are both WFFs, then the strings $(x \mathrel{\&} y)$, $(x \lor y)$, $(x \to y)$, and $(x \leftrightarrow y)$ are all WFFs.

So, now, give it a shot: How do you build $(P \lor Q) \to \sim R$?

From Rule 1, *P, Q,* and *R* are all WFFs:

P	*Q*	*R*	Rule 1

Then, from Rule 2, the string $\sim R$ is also a WFF:

P	*Q*	$\sim R$	Rule 2

Rule 3 says that the string $(P \lor Q)$ is a WFF:

$(P \lor Q)$ ~R Rule 3

And, then applying Rule 3 again, the string $((P \lor Q) \to \sim R)$ is a WFF.

$((P \lor Q) \to \sim R)$ Rule 3

Relaxing the rules

Technically speaking, every WFF in SL should begin and end with parentheses. But, in practice, this is rarely the case. As you can see throughout this book, statements rarely begin and end with parentheses.

Removing the outer parentheses from a WFF results in what logicians call the *relaxed* version of that WFF.

As you can see from the example in the previous section, the rules show that the string $((P \lor Q) \to \sim R)$ is a WFF. Is that WFF the same thing as $(P \lor Q) \to \sim R$? Strictly speaking, it isn't. So, Rule 4 — also known as the *relaxation rule* — takes care of this.

By convention, with this rule, you can remove the outer parentheses from a WFF to create a relaxed version of that WFF. Technically, though, the relaxed version of a WFF isn't a WFF, and Rule 4 isn't a rule. Instead, it's a convention that makes WFFs a little easier to read.

You may *not* use this relaxed version to build new WFFs.

Separating WFFs from non-WFFs

Remember that the purpose of these rules for building SL statements is not only to allow you to build WFFs, but also to prevent you from building strings that resemble WFFs (but really aren't). Take a look at the following messy string:

$\lor RQ \to) \sim (P$

You don't have much hope trying to build this wacky string. But, here's something that looks like it may be a WFF:

$(P \lor \sim Q) \to (R \lor S) \& \sim T$

Can you build this statement using the rules? Here's how you might try:

P	Q	R	S	T	Rule 1
P	$\sim Q$	R	S	$\sim T$	Rule 2
$(P \vee \sim Q)$	$(R \vee S)$	$\sim T$			Rule 3

At this point, you have all the parentheses you wanted to add, but you're still missing two operators. Even if you were to use the relaxed version of the statement, you can't get both operators into the statement without adding another set of parentheses, which is exactly the point. The string

$$(P \vee \sim Q) \rightarrow (R \vee S) \,\&\, \sim T$$

isn't a WFF. Its main operator could be either the $\rightarrow$-operator or the &-operator. And this ambiguity is a definite no-no because, as I cover in Chapter 4, an SL statement can have only one main operator.

Deciding whether a string is a WFF is, in a sense, like deciding whether a vessel is seaworthy. For instance, before setting a boat in the water, climbing in, and paddling away, you probably want to see whether it has any cracks or holes. If you do find some holes, you'd better patch them up (if they're fixable) *before* you set off.

What is true of boats in the water is also true of strings in SL. If you mistakenly think it's well-formed (seaworthy) and try to use it in a truth table or proof (sail it), you're going to run into trouble (sink like a stone).

Comparing SL to Boolean Algebra

In Chapter 13, I show you how you can get along in SL with only three operators ($\sim$, &, and $\vee$) and still do everything that you did with five operators.

The earliest version of formal logic — that is, logic with symbols rather than English words — made use of the fact that you can get along with only three operators. As I discuss in Chapter 2, Boolean algebra, invented by George Boole, was the first attempt at turning philosophical logic into a rigorous mathematical system. In fact, this form of logic was more closely linked to the mathematics that you learned in school than SL is.

Believe it or not, Boolean algebra is actually much easier than the algebra that you learned (or didn't) in school. For one thing, you work with only two numbers: 0 and 1. You also only have to worry about addition and multiplication. In this section, I show you the similarities between SL and Boolean algebra.

Reading the signs

Boolean algebra is really just a version of SL that uses different symbols, so in this section I start exploring Boolean algebra by making just a few little changes to SL.

In fact, I'm going to change only five symbols, which I outline in Table 14-1.

Table 14-1	Corresponding Symbols in SL and Boolean Algebra
SL Symbol	*Boolean Algebra Symbol*
T	1
F	0
~	−
&	×
∨	+

Constants are used in the same way in Boolean algebra as in SL, so when using the new symbols, you can draw the same basic truth tables that you're used to (see Chapter 6). For example, you can draw a truth table linking the &-operation in SL with Boolean multiplication:

P	Q	P & Q
T	T	T
T	F	F
F	T	F
F	F	F

P	Q	P × Q
1	1	1
1	0	0
0	1	0
0	0	0

Make sure you know that the &-operator in SL is multiplication (×) in Boolean algebra. Because & means *and*, you may mistakenly associate this operator with addition.

Notice that Boolean multiplication, symbolized by the times sign (×), is exactly the same as regular multiplication with 1s and 0s: $1 \times 1 = 1$ and anything multiplied by 0 is equal to 0.

Similarly, you can draw a truth table linking the ∨-operator in SL with Boolean addition:

P	Q	P ∨ Q
T	T	T
T	F	T
F	T	T
F	F	F

P	Q	P + Q
1	1	1
1	0	1
0	1	1
0	0	0

In this case, Boolean addition works like regular addition, but with one exception: In Boolean addition, $1 + 1 = 1$ because 1 is simply the highest number in the system.

Finally, you can draw a truth table linking the ~-operator with Boolean negation:

P	~P
T	F
F	T

P	−P
1	0
0	1

You may find it strange that in Boolean algebra $-1 = 0$ and $-0 = 1$. But remember that 1 and 0 in Boolean algebra are similar to **T** and **F** in SL. So, obviously it's not so strange to think of it as ~**T** = **F** and ~**F** = **T**.

Using the equal sign (=) with the other mathematical symbols is quite natural. Equality in Boolean algebra means the same thing as truth value in SL. You can also think of the equal sign as meaning that two statements are semantically equivalent, which is to say that they have the same truth value no matter what value the variables stand for (see Chapter 6 for more on semantic equivalence).

Table 14-2 shows the connection between the assignment of truth value in SL and equality in Boolean algebra.

Table 14-2	Using the Equal Sign (=) in Boolean Algebra
SL Truth Value or Semantic Equivalence	*Boolean Equality*
"The truth value of P is **T**."	$P = 1$
"The truth value of Q is **F**."	$Q = 0$
"The truth value of $P \lor {\sim}P$ is **T**."	$P + {-}P = 1$
"$P \& Q$ is semantically equivalent to $Q \& P$"	$P \times Q = Q \times P$
"${\sim}(P \lor Q)$ is semantically equivalent to ${\sim}P \& {\sim}Q$."	${-}(P + Q) = {-}P \times {-}Q$

Doing the math

In SL, you avoid writing strings that mix constants with the values **T** and **F**. In Boolean algebra, this avoidance isn't necessary. In fact, you can discover a lot about logic by mixing these different types of symbols. For example,

$$P \times 0 = 0$$

reminds you that any &-statement — remember, the multiplication sign stands in for the &-operator — that contains a false sub-statement (0) is false, no matter what the rest of the statement looks like. Similarly, the equation

$$P + 0 = P$$

tells you that when a $\lor$-statement contains a false sub-statement, its truth value depends on the value of the other sub-statement. And the equation

$$P \times 1 = P$$

tells you that when a &-statement contains a true sub-statement (1), its truth value depends upon the value of the other sub-statement.

Similarly, the equation

$$P + 1 = 1$$

tells you that when a $\lor$-statement contains a true sub-statement, the statement is always true.

Understanding rings and things

Both Boolean algebra and arithmetic in the non-negative integers (0, 1, 2, 3, and so on) are *semirings*, which means that they share a set of common properties.

For example, notice that the first three equations in the previous section are also true in regular arithmetic. They work simultaneously on two levels — as expressions of logical truth and as expressions of arithmetic truth.

In Chapter 10, I show you that in SL, the commutative, associative, and distributive properties from arithmetic carry over into logic. Table 14-3 shows a short list of important properties in Boolean algebra.

Table 14-3	Properties Common to Both Boolean Algebra and Arithmetic (and All Other Semirings)	
Property	**Addition**	**Multiplication**
Identity element	$P + 0 = P$	$P \times 1 = P$
Annihilator	n/a	$P \times 0 = 0$
Commutative	$P + Q = Q + P$	$P \times Q = Q \times P$
Associative	$(P + Q) + R = P + (Q + R)$	$(P \times Q) \times R = P \times (Q \times R)$
Distributive	$P \times (Q + R) = (P \times Q) + (P \times R)$	n/a

This set of properties is sufficient to classify both Boolean algebra and conventional arithmetic as examples of *semirings*. Every semiring has the five properties listed in Table 14-3.

Exploring syntax and semantics in Boolean algebra

Boolean algebra provides an interesting opportunity to focus on some of the differences between syntax and semantics that I discussed earlier in this chapter.

Boolean algebra and SL are quite different with regard to syntax, but very similar in terms of semantics.

On the level of syntax, Boolean algebra and SL are as different as French and German: Even if you understand one of them, you still have to learn the language if you want to understand the other.

But, after you know the syntactic rules of both, you can see that both systems can be used to express similar sorts of ideas. And looking at how symbols express ideas is the very heart of semantics. Notice how a statement in Boolean algebra has two separate meanings that are independent from each other. For example, consider the following statement:

$$1 \times 0 = 0$$

On one level, this statement expresses an arithmetic truth. In other words, when you multiply 1 by 0, the result is 0. But on another level, it expresses a logical truth, meaning that when you connect a true statement with a false statement using the word *and*, the result is a false statement. Depending on which semantic context you're working in, an equation in Boolean algebra works equally well to express a mathematical truth and a logical truth.

Part IV
Quantifier Logic (QL)

The 5th Wave   By Rich Tennant

WHILE DISSECTING HER ARGUMENT WITH UNIVERSAL INSTANTIATION, DEBORAH COUNTERS WITH HILBERT'S PROOF THEORY.

In this part . . .

In Part IV, you'll take a deeper cut into logic as you discover quantifier logic, or QL for short. QL uses everything you know about SL, but it extends this info to handle a wider variety of problems. In fact, QL is powerful enough to handle everything that all earlier forms of logic (2,000 years of them!) could do.

In Chapter 15, you get the basics of QL, including two new quantification operators. In Chapter 16, you find out how to translate statements from English into QL. In Chapter 17, I show you how to write proofs in QL using the skills that you already know from SL. In Chapter 18, I introduce relations and identities, which are two tools for making QL more expressive. Finally, in Chapter 19, you figure out how to use truth trees in QL, and you discover the surprising truth about infinite trees (how's that for a cliffhanger?).

Chapter 15

Expressing Quantity with Quality: Introducing Quantifier Logic

• •

• •

*I*n Chapter 3, I show you that logic is all about deciding whether an argument is valid or invalid. So, if you've read Chapter 3 and I say take a look at the following argument, you can probably tell me that it's a perfectly valid argument:

Premises:

> All of my children are honest.

> At least one of my children is a lawyer.

Conclusion:

> At least one honest lawyer exists.

This argument is indeed valid. However, the problem comes when you try to use sentential logic (SL) to show that it's valid. None of these three statements contains familiar words such as *and*, *or*, *if*, or *not*, so you can't use the five SL operators to express them.

So, the best you can do is to express the statements as SL constants. For example:

> Let *C* = All of my children are honest.

> Let *L* = At least one of my children is a lawyer.

> Let *H* = At least one honest lawyer exists.

After you express the statements as constants, you can put together the following argument:

C, L: H

Clearly, something important was lost in translation from English to SL, so no matter what method you use (truth table, quick table, truth tree, or proof), this SL argument is invalid. So, either the original argument really is invalid (which it isn't) or SL is wrong (which it isn't). Or, maybe you should peek at what's behind Door Number Three!

This chapter (as well as Chapters 16 through 19) gives you a peek behind that magic door. And behind that door, you find a solution to your problem. In this chapter, I help you understand why this argument is valid without completely trashing everything you've discovered about logic so far.

Instead, I show you how to take all of SL and extend it to a new and more powerful formal logical language called quantifier logic. This newfangled language introduces you to two new symbols that allow you to express logical concepts that are way beyond the reach of SL. Get ready because a whole new logical world awaits!

Taking a Quick Look at Quantifier Logic

Quantifier logic (QL) uses much of the same structure as sentential logic (SL). You may recall that SL has both constants and variables (see Chapter 4). A constant represents a statement in a natural language (such as English) whereas a variable represents a statement in SL. QL has both constants and variables, but it uses them in a different way that allows you to break statements into smaller pieces for greater precision.

In some books, quantifier logic is also called *quantificational logic, predicate logic,* or *first-order predicate logic.*

Using individual constants and property constants

QL has the following two types of constants (rather than just one like in SL):

> ✔ **Individual constants:** An individual constant represents the *subject* of a sentence with a *lowercase letter* from *a* to *u*.
>
> ✔ **Property constants:** A property constant represents the *predicate* of a sentence with a *capital letter* from *A* to *Z*.

As in SL, the formal logical language QL allows you to translate statements in English into symbols. You may remember from SL that when you want to translate a statement from English into formal logic, the first step is to define some constants. For example, suppose you want to translate the following statement:

David is happy.

First, define an individual constant to represent the subject of the sentence:

Let *d* = David

Next, define a property constant to represent the predicate of the sentence:

Let *Hx* = *x* is happy

To write the statement in QL, just replace the *x* with the individual constant:

Hd

In English, the subject usually comes before the predicate. But in QL, this order is always reversed. This reversal can be confusing at first, so make sure you keep it clear.

You can already see from this example the added flexibility QL affords you. With SL, you'd only be able to assign one variable to the entire statement, like this:

Let *D* = David is happy.   (SL TRANSLATION)

Here's another example of how to translate an English statement into QL:

The cat was on the mat.

First define both the individual and the property constants:

Let *c* = the cat

Let *Mx* = *x* was on the mat

Now, you can translate the original statement as

> *Mc*

Note that this translation works well no matter how long the subject and predicate are. For example, you can even translate the following verbose sentence:

> My obstreperous aunt from Wisconsin with three sniveling French poodles is leaving on a three-month tour of six countries in Europe and Africa.

In this case, define your constants as follows:

> Let *a* = my obstreperous aunt from Wisconsin with three sniveling French poodles
>
> Let *Lx* = *x* is leaving on a three-month tour of six countries in Europe and Africa

Then, you can easily translate the statement as

> *La*

Only use an individual constant to represent a single thing — never more than one thing.

For instance, when looking at the previous example, the subject of the sentence is the aunt, and the poodles are just part of her description. But suppose the sentence read like this:

> My obstreperous aunt from Wisconsin *and* her three sniveling French poodles *are* leaving on a three-month tour of six countries in Europe and Africa.

In QL, you can't use one constant to represent the group of three poodles. Instead, you need to define three more individual constants — one for each of the three poodles. For example, suppose the poodles are named Fifi, Kiki, and Elvis. Then you might use the following definitions:

> Let *f* = Fifi
>
> Let *k* = Kiki
>
> Let e = Elvis

Then, because they're all going on tour together, you could express the statement as

La & Lf & Lk & Le

As you can see, the &-operator from SL is also used in QL. In the next section, you see how all five SL operators come into play.

Formally defining constants is an important technical point, but it's easy and you'll probably get the hang of it quickly. So, from now on, I leave it out unless it's necessary for clarity.

Incorporating the SL operators

QL contains the five operators from SL: ~, &, ∨, →, and ↔. You can get a lot of mileage out of this new formulation for statements, especially when you add in the SL operators that you already know. Suddenly, with QL, you can translate sentences into symbols at a much deeper level than you could when just using SL.

For example, take the statement "Nadine is an accountant and her secretary is on vacation." In SL, you can only crudely break the sentence into two parts:

N & S   (SL TRANSLATION)

With QL, however, you can distinguish four separate pieces of the sentence and use these four pieces to represent the sentence as follows:

An & Ms

Similarly, the statement

Genevieve is either a Scorpio or an Aquarius.

can be translated as

Sg ∨ Ag

Here, the sentence is understood as "*Either* Genevieve is a Scorpio *or* Genevieve is an Aquarius."

In a similar way, the statement

If Henry is a nice guy then I am the Queen of Sheba.

can be translated as

$Nh \rightarrow Qi$

And, finally, the statement

Five is odd if and only if it's not divisible by two.

can be translated as

$Of \leftrightarrow \sim Df$

In this sentence, the word *it* is a pronoun that stands for *five*, so I use the individual constant *f* in both cases.

As you can see from these examples, the single-letter constants of SL have been replaced by double-letter terms in QL. But, never fear, you can still combine these terms according to the same rules you used in SL.

Understanding individual variables

In QL, an *individual variable* represents an *unspecified* subject of a sentence with a lowercase letter *v, w, x, y,* or *z*.

You've already seen individual variables used in the formal definition of property constants. For example, in the definition

Let $Hx = x$ is happy

the individual variable is *x*. And as you've already seen, you can replace this variable with an individual constant to create a QL statement. For example, when you represent the sentence "David is happy" in QL as

Hd

you replace the variable *x* with the constant *d*.

You may notice that QL has individual constants, individual variables, and property constants, but not *property variables*. Property variables come into play in *second-order predicate logic* (see Chapter 21), which is why QL is also called first-order predicate logic.

To save space, I shorten the phrase *individual variable* to *variable* throughout Part IV.

Expressing Quantity with Two New Operators

So far in this chapter, you've looked at a few examples to see how QL represents statements that *can* be expressed in SL. Now, in this section, you're going to see how QL goes into territory that SL *can't* cover.

The main vehicles for this new expressiveness are two new symbols: ∀ and ∃.

Understanding the universal quantifier

The symbol ∀ is called the *universal quantifier*. It's always used with a variable, such as *x*, with ∀*x* being translated into English as "For all *x* . . ."

At the beginning of this chapter, you saw how troublesome it was to translate the statement "All of my children are honest" into SL in any useful way. But with the universal quantifier, you now have the power.

For example, first you need to attach the universal quantifier to a variable. Set up the problem with empty brackets, as shown here:

∀*x* [ ]

After you've set this problem up, you can use the rules you already know to translate the rest of the statement. Then you place the result in the brackets:

∀*x* [*Cx* → *Hx*]

You read this statement as "For all *x*, if *x* is my child, then *x* is honest." It may seem a bit clunky at first, but after you get used to the setup, the possibilities are literally endless.

You can also use ∀ to translate words like *every* and *each*. I get into translations in detail in Chapter 16.

Expressing existence

The symbol ∃ is called the *existential quantifier*. It's always used with a variable, such as *x*, with ∃*x* being translated into English as "There exists an *x* such that . . ."

At the beginning of this chapter, you saw that SL is limited in its ability to express the statement "At least one of my children is a lawyer." However, the existential quantifier makes expression of this statement possible.

First, as with the universal quantifier, attach the existential quantifier to a variable:

∃*x* [ ]

Now, inside the brackets, translate the rest of the statement:

∃*x* [*Cx* & *Lx*]

You can read this statement as "There exists an *x* such that *x* is my child and *x* is a lawyer."

Similarly, you can use ∃ to translate the statement "At least one honest lawyer exists:"

∃*x* [*Hx* & *Lx*]

This statement translates literally "There exists an *x* such that *x* is a lawyer and *x* is honest."

So, here's your first argument in QL:

∀*x* [*Cx* → *Hx*], ∃*x* [*Cx* & *Lx*] : ∃*x* [*Hx* & *Lx*]

As you can see, a lot more is going on here than in the SL version. And that's a good thing, because QL has captured in symbols exactly what's valid about this argument. But, before you look at how to prove that this argument is valid, you'll need to get a little more background.

You can also use ∃ to translate words such as *some*, *there is*, and *there are*. I discuss translations of this kind in Chapter 16.

Creating context with the domain of discourse

Because QL can handle a wider range of circumstances than SL, a new type of ambiguity arises when translating from English to QL.

For example, look at this statement:

Everyone is wearing a suit.

You can translate this statement into QL as follows:

$\forall x\, [Sx]$

This translates literally as "For all x, x is wearing a suit," which is false. But if the context you are speaking of is a meeting of four male coworkers, the statement might be true.

To prevent this potential for confusion, you need to be clear about the context of a statement. For this reason, you must place variables in a context called the *domain of discourse*.

The domain of discourse provides a context for statements that contain variables. For example, if the domain of discourse is *people*, then x can stand for me, or you, or your Aunt Henrietta, but not for the Bart Simpson, your cell phone, or the planet Jupiter.

In math, the domain of discourse is often taken for granted. For example, when you write the equation $x + 2 = 5$, you can assume that the variable x stands for a number, so the domain of discourse in this case is the set of all numbers.

How you translate a statement from English into QL is often dependent upon the domain of discourse. To see why this is so, take a look at two translations of an argument using two different domains of discourse:

Premises:

All of my children are honest.

At least one of my children is a lawyer.

Conclusion:

At least one honest lawyer exists.

For the first translation, here's a declaration I use:

Domain: Unrestricted

An unrestricted domain means that x can be literally anything: a flea, a unicorn, the Declaration of Independence, the Andromeda Galaxy, the word *butter*, or your first kiss. (You thought I was kidding when I said *anything*, didn't you?) This wide range isn't a problem, though, because the statements themselves put restrictions on what x can be.

With this domain of discourse, the argument translates into QL as

$$\forall x \, [Cx \rightarrow Hx], \exists x \, [Cx \,\&\, Lx] : \exists x \, [Hx \,\&\, Lx]$$

You may think that the domain of discourse is just a technical point that doesn't really affect the argument, but that isn't the case. In fact, when writing an argument, you can use the domain of discourse to your advantage by cleverly choosing a domain that makes the translation from English into QL easier and more concise.

For this second translation of the same argument, I can start out by declaring a different domain:

Domain: My children

Now, all of the variables are under a significant restriction, which means that I can translate the statement "All of my children are honest" as

$$\forall x \, [Hx]$$

instead of the earlier statement I used:

$$\forall x \, [Cx \rightarrow Hx]$$

You read $\forall x[Hx]$ as "For all x, x is honest" or simply, "All x is honest." Note that the statement itself doesn't mention the fact that x is one of my children (Cx from the previous statement) because the domain of discourse handles this restriction.

Similarly, using the same domain of discourse, you can translate the statement "At least one of my children is a lawyer" in the following way:

$$\exists x \, [Lx]$$

You read this statement as "There exists an x such that x is a lawyer." Again, the statement itself doesn't mention that x is one of my children because this restriction is covered by the domain of discourse.

Finally, you can translate "At least one honest lawyer exists" as

$\exists x\,[Hx \,\&\, Lx]$

This is the same translation as when the domain is unrestricted because the statement contains no mention of my children. Thus, within the domain "My children," the same argument translates as

$\forall x\,[Hx],\ \exists x\,[Lx] : \exists x\,[Hx \,\&\, Lx]$

As you can see, this argument is a little shorter and simpler than the argument using an unrestricted domain.

Be sure to use just *one* domain of discourse for a single argument or for any statements that you're analyzing as a group. Switching between domains is a lot like trying to add apples and oranges, and the result is usually a mixed up argument that doesn't give you proper results.

The domain of discourse is important from a technical standpoint, but most of the time you can use an unrestricted domain. So, from now on, unless I state otherwise, assume that the domain is unrestricted.

Picking out Statements and Statement Forms

In Chapter 5, I describe the difference in SL between a statement, for example, $(P \,\&\, Q) \to R$, and a statement form, such as $(x \,\&\, y) \to z$. Simply put, a statement has constants but no variables, and a statement form has variables but no constants.

In QL, however, a single expression can have both constants and variables. In this section, I explain how constants and variables get used, and show you what constitutes a statement in QL.

Determining the scope of a quantifier

In Chapter 5, I discuss how parentheses limit the scope of an SL operator. Consider, for example, this SL statement:

$$P \,\&\, (Q \vee R)$$

The &-operator in this statement is *outside* of the parentheses, so its scope is the whole statement, which means that this operator affects every variable in the statement. On the other hand, the ∨-operator is *inside* the parentheses, so its scope is limited to what's inside the parentheses. That is, this operator affects only the constants Q and R, but not P.

In QL, the scope of the quantifiers ∀ and ∃ is limited in a similar way, but with brackets instead of parentheses. For example, check out this statement:

$$Ax \,[Cx \rightarrow Nx] \,\&\, Rb$$

The scope of the quantifier ∀ is limited to what's inside the brackets: $Cx \rightarrow Nx$. In other words, this quantifier doesn't affect the &-operator or the sub-statement Rb.

Discovering bound variables and free variables

You thought you knew enough about variables already, didn't you? Well, get ready because here's another exciting lesson. When you're sitting at your desk pondering an expression, remember that every variable in that expression is considered either *bound* or *free*. Here's what I mean:

- ✔ A *bound variable* is a variable that's being quantified.

- ✔ A *free variable* is a variable that isn't being quantified.

Easy enough, right? To completely understand the difference between bound variables and free variables, take a look at the following expression:

$$\exists x \,[Dx \rightarrow My] \,\&\, Lx$$

Can you guess which of the variables in this expression are bound and which are free?

When a variable is *outside* of all brackets, it's always free. This is because the scope of a quantifier is always limited to what's *inside* a set of brackets. So, the x in Lx is a free variable.

However, also remember that just because a variable is inside a set of brackets doesn't mean that it's bound. In the example, the variable y is also free even though it's inside the brackets. This is because these brackets define the scope of $\exists x$, which is operating on x, not y.

You can probably guess by now that the variable x in Dx is a bound variable. You can be sure of this because it appears inside the brackets and because these brackets define the scope of $\exists x$, which does quantify x.

The difference between bound variables and free variables becomes important in Chapter 17 when you dive into QL proofs. For now, make sure you understand that a bound variable is being quantified and a free variable isn't being quantified.

Knowing the difference between statements and statement forms

With your new knowledge about bound variables and free variables from the previous section, you'll easily be able to tell the difference between statements and statement forms.

When you're evaluating an expression, you can be sure that each one falls into one of two categories:

- ✔ **Statements:** A *statement* contains *no* free variables.
- ✔ **Statement forms:** A *statement form* contains at least *one* free variable.

Distinguishing a statement from a statement form is that easy. For example, consider the following expression:

$$\forall x \, [Cx \rightarrow Hx] \ \& \ Ok$$

This expression is a statement because it has no free variables. Note that both instances of the variable x appear inside the brackets within the scope of the quantifier $\forall x$. Additionally, k appears outside the brackets, but it's a constant, not a variable.

On the other hand, consider this expression:

$$\forall x \; [Cx \rightarrow Hx] \; \& \; Ox$$

This expression has one free variable (the x in Ox), which falls outside of the brackets and is, therefore, outside the scope of the quantifier $\forall x$. So, the expression is a statement form rather than a statement.

Formal definitions in QL use statement forms that you can then turn into statements. For example, look at this definition:

Let $Ix = x$ is Italian

The expression Ix is a statement form because the variable x is free. But, you turn this statement form into a statement when you replace the variable with an individual constant. For example, to translate the statement

Anna is Italian.

you might write:

Ia

Similarly, binding the variable in a statement form also turns the statement form into a statement. For example, to translate the statement

Somebody is Italian.

you can write

$\exists x \; [Ix]$

The difference between statements and statement forms becomes important in Chapter 18 when you begin writing QL proofs. For now, just make sure you understand that a statement has no free variables.

Chapter 16

QL Translations

• •

In This Chapter

▶ Recognizing the four basic forms of categorical statements as *all*, *some*, *not all*, and *no* statements

▶ Using either the ∀ or the ∃ quantifier to translate each of the basic forms

▶ Translating English statements that begin with words other than *all*, *some*, *not all*, or *no*

• •

*I*n Chapter 15, I introduce you to quantifier logic (QL), paying special attention to the two quantifiers ∀ and ∃. I also show you how to translate a few simple statements into QL.

In this chapter, I show you how to translate the four most basic forms of *categorical statements* from English into QL. These tend to be statements that begin with the words *all*, *some*, *not all*, and *no*. (Take a quick look at Chapter 2 for more information about categorical statements.)

This chapter also covers how to translate each of these four forms using either the ∀ or the ∃ quantifier. Finally, I show you how to recognize categorical statements that don't begin with any of the four key words.

Translating the Four Basic Forms of Categorical Statements

In this section, I show you how to translate the words *all, some, not all,* and *no.* You'll use these forms a lot while learning to translate from English into QL. (If you need to, take a peek at Chapter 2 to make sure you understand the basic forms.) Next, I show you how to translate statements beginning with *all* and *some.* After you understand how to translate those statements, the statements beginning with *not all* and *no* are pretty easy.

After you know how to translate statements beginning with the words *all*, *some*, *not all*, and *no*, a lot of the hard work is done. In general, the easiest way to approach these types of statements is to set up a quantifier and brackets as in Table 16-1:

Table 16-1	Translations of the Four Basic Forms of Categorical Statements
English Word	*QL Quantifier*
All	∀ [ ]
Some	∃ [ ]
Not all	~∀x [ ]
No	~∃x [ ]

"All" and "some"

Translating statements from English into QL isn't difficult, but you have to be aware of a couple of sticking points regarding restricted and unrestricted domains. (For a refresher on how to use the domain of discourse, flip to Chapter 15.)

In this section, I show you how to translate statements beginning with the words *all* and *some* into QL.

Using a restricted domain

Two statements in English may be identical except for the words *all* or *some*. For this reason, you may think that their QL translations would be identical except for the choice of quantifier (∀ or ∃). When the domain is restricted, this is often exactly how things work out.

For example, take the following two statements:

> All horses are brown.

> Some horses are brown.

Suppose you use a restricted domain of discourse:

> Domain: Horses

If you use a restricted domain, you need to define only one property constant to translate the two statements. For example:

Let Bx = x is brown

Now you can translate both statements as follows:

$\forall x \, [Bx]$

$\exists x \, [Bx]$

As you can see, the only difference between the two QL statements is that you replaced $\forall$ in the first statement with $\exists$ in the second. Just as you'd expect, nothing tricky here.

For the rest of this chapter, I focus on translations in an unrestricted domain.

Using an unrestricted domain

Restricting the domain isn't always possible, so this time, suppose that your domain of discourse is unrestricted:

Domain: Unrestricted

Now, to translate the statements from the previous section, you need another property constant. So, here are the two constants you need, including the one from the previous section:

Let Bx = x is brown

Let Hx = x is a horse

Let Fx = x can fly

Now you can translate the statement "All horses are brown," as

$\forall x \, [Hx \rightarrow Bx]$

This statement translates literally as "For all x, if x is a horse, then x is brown."

But, on the other hand, you translate the statement "Some horses are brown," as

$\exists x \, [Hx \, \& \, Bx]$

This statement translates literally as "There exists an x such that x is a horse and x is brown."

Notice that the difference between the two statements is more than just the quantifiers $\forall$ and $\exists$. In the first statement, you use a $\rightarrow$-operator, whereas in the second you use a &-operator.

Take a look at what happens, though, if you mix up these operators. Suppose you tried to translate the first statement as

$\forall x \ [Hx \ \& \ Bx]$ WRONG!

This translation is wrong because what it actually says is, "For all x, x is both a horse and brown." More informally, it says "Everything is a brown horse." And because the domain is unrestricted, everything really does mean *everything*, which obviously isn't what you meant.

As another example, suppose that you tried to translate the statement

Some horses can fly.

as

$\exists x \ [Hx \rightarrow Fx]$ WRONG!

Again, this translation is wrong because it says "There exists an x such that if x is a horse, then x can fly." The problem here is a bit more subtle.

Suppose x is something other than a horse — for example, a shoe. Then, the first part of the statement inside the brackets is false, which makes everything inside the brackets true. So, in this case, the statement seems to be telling you that the existence of a shoe means that some horses can fly. Again, something has gone wrong with the translation.

When translating from English into SL, use the following rules of thumb:

- When using $\forall$, use a $\rightarrow$-statement inside the brackets.
- When using $\exists$, use a &-statement inside the brackets.

"Not all" and "no"

After you know how to translate the two positive forms of categorical statements, translating the two negative forms — *not all* and *no* — is easy. As I discuss in Chapter 2, a *not all* statement is just the contradictory form of an *all* statement. Similarly, a *no* statement is the contradictory form of a *some* statement.

For example, you already know how to translate the statement

All horses are brown.

as

$\forall x\ [Hx \rightarrow Bx]$

Thus, its contradictory form is

Not all horses are brown.

This statement translates easily as follows:

$\sim\forall x\ [Hx \rightarrow Bx]$

Furthermore, because *all* and *not all* statements are contradictory, you know that exactly one of these statements is true and the other is false.

Similarly, you also know how to translate the statement

Some horses can fly.

as

$\exists x\ [Hx\ \&\ Fx]$

This statement's contradictory form is

No horses can fly.

As you may have guessed, this statement translates as

$\sim\exists x\ [Hx\ \&\ Fx]$

Again, because *some* and *no* statements are contradictory, exactly one is true and the other is false.

Discovering Alternative Translations of Basic Forms

QL is very versatile and offers you more than one way to translate each form. In this section, I show you how to translate each of the four basic forms of categorical statements using both the ∀ and ∃ quantifiers.

Knowing both versions of each translation will help you understand the hidden connection between the two quantifiers ∀ and ∃.

For the examples in this section, I use the following definitions:

Let Dx = x is a dog

Let Px = x is playful

Table 16-2 organizes the information from this section about translating the four basic forms from English into QL. For each form, I list two English statements. The first statement gives the simplest way to say it in English and its most direct translation into QL. The second gives an alternative phrasing of the same idea with a translation of this phrasing.

Table 16-2	Alternative Translations of the Four Basic Forms of Categorical Statements
English Translations	**QL Translations**
All dogs are playful.	$∀ [Dx → Px]$
(No dogs are not playful.)	$(∼∃ [Dx \& ∼Px])$
Some dogs are playful.	$∃ [Dx \& Px]$
(Not all dogs are not playful.)	$(∼∀ [Dx → ∼Px])$
Not all dogs are playful.	$∼∀ [Dx → Px]$
(Some dogs are not playful.)	$(∃ [Dx \& ∼Px])$
No dogs are playful.	$∼∃ [Dx \& Px]$
(All dogs are not playful.)	$(∀ [Dx → ∼Px])$

Translating "all" with ∃

It makes perfect sense to translate the statement "All dogs are playful" as

$\forall x\ [Dx \to Px]$

Think about it. Even if it is upside down, the ∀ stands for *all*. That's why I discussed this translation first.

But, when you think about it some more, you may realize that when you say all dogs are playful, you're making an equivalent statement:

No dogs are *not* playful.

In other words, this statement is just a *no* statement with the word *not* inserted. So, a perfectly good translation is:

$\sim\!\exists x\ [Dx\ \&\ \sim\!Px]$

The literal translation here is this: "It isn't true that there exists an *x* such that *x* is a dog and *x* is not playful." But, both English statements mean the same thing, so both QL statements are semantically equivalent. (For more on semantic equivalence, see Chapter 6.).

Translating "some" with ∀

In the same way that you can translate *all* with ∃, you can use the ∀ quantifier to translate the word *some*. For example, consider the statement

Some dogs are playful.

The direct translation of this statement into QL is

$\exists x\ [Dx\ \&\ Px]$

In this case, note that this statement means the same thing as

Not all dogs are *not* playful.

So, you can treat this statement as a *not all* statement with the word *not* inserted. It translates as

$\sim\!\forall x\ [Dx \to \sim\!Px]$

In this case, the literal translation is "It isn't true that for all x, if x is a dog, then x is not playful." But, the two statements in English mean the same thing, so the two QL statements are semantically equivalent.

Translating "not all" with ∃

Suppose you want to translate the following statement:

Not all dogs are playful.

You already know how to translate this statement as

~∀x [$Dx → Px$]

But, here's another way to think about it: Because not all dogs are playful, *there's at least one* dog somewhere that is *not* playful. So, you can rephrase the original statement as

Some dogs are not playful.

To express this idea, you can use the following translation:

∃x [Dx & ~Px]

The more exact translation for this statement is "There exists an x such that x is a dog and x is not playful." But these two English statements mean the same thing, and so do their counterparts in QL.

Translating "no" with ∀

Suppose you want to translate this statement:

No dogs are playful.

The simplest translation is

~∃x [Dx & Px]

The more exact translation here is "It is not true that there exists an x such that x is a dog and x is playful."

As you may have guessed, you can think of this translation another way. For instance, the fact that no dogs are playful means the following:

All dogs are not playful.

You can express this idea in QL as

$\forall x\, [Dx \rightarrow \sim Px]$

The more exact translation in this case is: "For all x, if x is a dog, then x is not playful."

Once again, because the two English statements mean the same thing, their translations into QL are semantically equivalent.

Identifying Statements in Disguise

So, now you understand how to translate the four most basic forms of categorical statements from English into QL. Well, you're in for a treat because in this section, I help you broaden this understanding.

Here, I show you how to recognize statements that don't start with the exact words you are used to. When you're done, you'll be ready to take on a much wider variety of statements in English and turn them into QL.

Recognizing "all" statements

Even if a statement doesn't begins with the word *all*, it may mean something close enough to use the $\forall$ quantifier. Here are some examples, followed by translations into QL:

Any father would risk his life to save his child.

$\forall x\, [Fx \rightarrow Rx]$

Every man in this room is single and eligible.

$\forall x\, [Mx \rightarrow (Sx \,\&\, Ex)]$

The family that prays together stays together; while every family that takes vacations together always laughs a lot.

$\forall x\, [Px \rightarrow Sx] \,\&\, \forall x\, [Vx \rightarrow Lx]$

In the last example, you can quantify the variable *x* twice because you're making a statement about *all* families.

Recognizing "some" statements

You may run across certain statements that are close enough to *some* statements that you'll want to use the ∃ quantifier to translate them. For example, consider these statements with their translations:

At least one of the guests at this party committed the crime.

∃x [Gx & Cx]

Many teenagers are rebellious and headstrong.

∃x [Tx & (Rx & Hx)]

Both corrupt plumbers and honest judges exist.

∃x [Px & Cx] & ∃y [Jy & Hy]

Notice in the last example that you need two variables — *x* and *y*. These variables make it clear that while both groups (corrupt plumbers and honest judges) exist, they don't necessarily overlap.

Recognizing "not all" statements

You can easily translate a few statements that don't include the words *not all* by using ~∀. For example, consider these statements that are considered *not all* statements:

Not every musician is flaky.

~∀x [Mx → Fx]

Fewer than 100 percent of accountants are reserved and thorough.

~∀x [Ax → (Rx & Tx)]

All that glitters is not gold.

~∀x [Ix → Ox] (Because the words *glitters* and *gold* begin with the letter *g*, I'm using the first vowel in each case.)

Even though this last example begins with the word *all*, it's a *not all* statement that means "Not all things that glitter are gold." So, be sure to think through what a statement is really saying before translating it!

Recognizing "no" statements

You've probably guessed by now that some statements that don't start with the word *no* are easy to translate using ~∃. You're right on the money. Consider these statements and their translations:

Not a single person on the jury voted to acquit the defendant.

~∃x [Jx & Vx]

There aren't any doctors or even college graduates in our family.

~∃x [Fx & (Dx ∨ Cx)]

Nobody who buys a restaurant makes money without putting in long hours.

~∃x [Bx & (Mx & ~Px)]

You can think of the last statement as "*No* person exists who buys a restaurant *and* makes money *and* doesn't put in long hours."

Chapter 17

Proving Arguments with QL

- -

In This Chapter

▶ Comparing proofs in SL and QL

▶ Applying quantifier negation (**QN**) in QL

▶ Using the four quantifier rules **UI**, **EI**, **UG**, and **EG**

- -

*H*ere's the good news: If you've got the hang of sentential logic (SL) proofs, you already know 80 percent of what you need to write proofs in quantifier logic (QL). So, first you need to figure out how much you already know. If you need a quick refresher, take a look at the chapters in Part III, which tell you everything you need to know.

In this chapter, I start out by showing you how proofs in QL are similar to those in SL. One way that they're similar is that they both use the eight implication rules and the ten equivalence rules. I show you exactly how and when to use these rules.

After you get comfortable with a few simple proofs in QL, I also introduce an additional rule, quantifier negation (**QN**), which allows you to make changes that affect the quantifier (∀ or ∃). Luckily, this rule, which is the first rule that's unique to QL, is an easy one to master.

The rest of the chapter focuses on the four quantifier rules: Universal Instantiation (**UI**), Existential Instantiation (**EI**), Universal Generalization (**UG**), and Existential Generalization (**EG**). The first two rules allow you to remove each type of quantifier from a QL statement. The other two rules allow you to add either type of quantifier to a QL statement.

As a group, the four quantifier rules allow you to use the eight implication rules from SL to their maximum benefit. Read on to find out how.

Applying SL Rules in QL

The 18 rules of inference from SL are one huge slice of the 80 percent you already know about writing proofs in QL. These rules include the eight implication rules from Chapter 9 and the ten equivalence rules from Chapter 10. In smany cases, you can transfer these rules over to QL proofs with only minor adjustments.

This section shows you the many similarities between proofs in QL and SL.

Comparing similar SL and QL statements

One key difference between SL and QL is the way the two languages handle simple statements. For example, to translate the statement "Howard is sleeping" into SL, you might write:

H (SL TRANSLATION)

To translate the same statement into QL, you might write:

Sh (QL TRANSLATION)

Similarly, to translate the more complex statement "Howard is sleeping and Emma is awake" into SL, you might write:

H & E (SL TRANSLATION)

To translate the same statement into QL, you might write:

Sh & Ae (QL TRANSLATION)

Neither of these English statements contains any words that require you to use a quantifier (∀ or ∃). That's why you can translate them both into either SL or QL.

In fact, in such cases only one difference exists between SL and QL translations: the constants. For example, in SL, you translate a simple English statement into a single-letter constant (for example, *H* or *E*), which can then stand on its own as an SL statement.

In QL, however, you translate the same English statement into a two-letter combination of a property constant and an individual constant (for example,

Sh or *Ae*), which can then stand on its own as a QL statement. (Flip to Chapter 4 for more on SL constants and Chapter 15 for QL constants.)

For QL statements, the rule of thumb is just to treat the two-letter combinations as *indivisible* building blocks of larger statements. That is, you're not going to break them apart, so you can just treat them as units, as you would treat single-letter constants in SL. (See Chapter 18 to learn about identities, the one exception to this rule.)

Transferring the eight implication rules from SL into QL

Because SL and QL are so similar, you can easily apply the eight implication rules to any QL statement or statement form, just as you would in SL.

As in SL, when you use the eight implication rules in QL, you can only apply them to whole statements or statement forms — never to partial statements or statement forms. (Check out Chapter 9 for a refresher on how to use the implication rules.)

Working with QL statements without a quantifier

Here I show you how the eight implication rules apply to QL statements without a quantifier. For example, suppose you want to prove this argument:

$Jn \rightarrow Bn$, ~$Bn \lor Ep$, Jn : Ep

Begin, as always, by listing the premises:

1.	$Jn \rightarrow Bn$	**P**
2.	~$Bn \lor Ep$	**P**
3.	Jn	**P**

Now, continue as you would with an SL proof, but treat the two-letter constants as indivisible chunks, just like the single-letter constants and variables in QL:

4.	Bn	1, 3 **MP**
5.	Ep	2, 4 **DS**

Working with QL statements with a quantifier

Using the eight implication rules from SL also works for QL statements with a quantifier. As I noted in the preceding section, just make sure you're applying the rules to whole statements and statement forms.

For example, suppose you want to prove the following argument:

$\forall x\ [Nx]\ \&\ \exists y\ [Py] : (\forall x\ [Nx]\ \lor\ \forall y\ [Py])\ \&\ (\exists x\ [Nx]\ \lor\ \exists y\ [Py])$

Again, start with the premise:

1.	$\forall x\ [Nx]\ \&\ \exists y\ [Py]$	**P**

Now you can work with this entire statement as in SL:

2.	$\forall x\ [Nx]$	1 **Simp**
3.	$\exists y\ [Py]$	1 **Simp**
4.	$\forall x\ [Nx]\ \lor\ \forall y\ [Py]$	2 **Add**
5.	$\exists x\ [Nx]\ \lor\ \exists y\ [Py]$	3 **Add**
6.	$(\forall x\ [Nx]\ \lor\ \forall y\ [Py])\ \&\ (\exists x\ [Nx]\ \lor\ \exists y\ [Py])$	4, 5 **Conj**

You can't apply the implication rules to part of a statement, even when that part is the only thing in the brackets of a statement with a quantifier.

For example, here's an *invalid* argument:

Premises:

Some members of my family are doctors.

Some members of my family never finished high school.

Conclusion:

Some doctors never finished high school.

And here's the "proof" of this argument:

$\exists x\ [Fx\ \&\ Dx],\ \exists x\ [Fx\ \&\ Nx] : \exists x\ [Dx\ \&\ Nx]$

1.	$\exists x\ [Fx\ \&\ Dx]$	**P**	
2.	$\exists x\ [Fx\ \&\ Nx]$	**P**	
3.	$\exists x\ [Dx]$	1 **Simp**	WRONG!
4.	$\exists x\ [Nx]$	2 **Simp**	WRONG!
5.	$\exists x\ [Dx\ \&\ Nx]$	3, 4 **Conj**	WRONG!

Obviously, something is very wrong here. The moral of the story is that you aren't allowed to apply **Simp**, **Conj**, or any other implication rule to the stuff inside the brackets and just ignore the rest of the statement. (Later in this chapter, I show you how to prove arguments like this one without breaking this rule.)

Working with QL statement forms

In Chapter 15, I show you that in QL, a property constant combined with an individual variable (for example, *Px*) is not really a statement unless it's inside a pair of brackets and modified by a quantifier (for example, ∀*x* [*Px*]). When it appears free, it isn't a statement at all, but rather a *statement form*.

The good news is that when writing proofs, you can treat statement forms the same as statements. Later in the chapter, I give specific examples of how statement forms can arise in the course of a QL proof. For now, just know that the eight implication rules also apply to statement forms.

Employing the ten SL equivalence rules in QL

You can apply the ten equivalence rules to any *whole* QL statement or statement form or to any *part* of a QL statement or statement form.

In Chapter 10, I explain that the equivalence rules give you greater flexibility than the implication rules. One reason for this flexibility is that you can apply the ten equivalence rules not only to whole SL statements, but also to parts of statements. This same application also works in QL. For example, suppose you want to prove the following argument:

∃*x* [*Cx* → ~*Sx*] : ∃*x* ~[*Cx* & ~*Sx*]

1.   ∃*x* [*Cx* → ~*Sx*]   **P**

Because the ∃*x* remains unchanged from the premise to the conclusion, your main challenge here is to change the rest of the statement. And you can do this by using two of your favorite equivalence rules:

2.   ∃*x* [~*Cx* ∨ ~*Sx*]   1 **Impl**

3.   ∃*x* ~[*Cx* & *Sx*]   2 **DeM**

This flexibility with regard to equivalence rules also applies equally well to statement forms.

For example, the expression $Cx \to {\sim}Sx$ is a statement form because the variable x is unquantified (see Chapter 15). You can use the equivalence rule **Impl** (see Chapter 10) to rewrite this statement form as ${\sim}Cx \lor {\sim}Sx$.

Later in the chapter, I discuss how sentence forms find their way into QL proofs.

Transforming Statements with Quantifier Negation (QN)

In transferring over the rules for SL proofs into QL, you've had to work around the quantifiers $\forall$ and $\exists$. For some proofs, working around the quantifiers in this way won't hold you back, but suppose you want to prove this argument:

$$\forall x\,[Hx \to Bx] : {\sim}\exists x\,[Hx \,\&\, {\sim}Bx]$$

This argument may look familiar if you've read Chapter 16. There, I use the premise statement to stand for "All horses are brown" and the conclusion statement to stand for "No horses are not brown." Because these two statements are equivalent, you can start with the first and prove the second.

The problem here is that the quantifier changes from $\forall$ to ${\sim}\exists$, and nothing from SL is going to help with that. Luckily, **QN** comes to the rescue. Read on to see how.

Introducing QN

Quantifier Negation (**QN**) allows you to change a QL statement to an equivalent statement by taking the following three steps:

1. **Place a ~-operator immediately before the quantifier.**

2. **Change the quantifier (from $\forall$ to $\exists$, or from $\exists$ to $\forall$).**

3. **Place a ~-operator immediately after the quantifier.**

As in SL, if the resulting statement has any instances of two adjacent ~-operators, you can remove both of them. This removal uses the double negation rule (**DN**) without referencing it, as I discuss in Chapter 10.

Table 17-1 lists the four distinct cases in which you can use **QN**.

Table 17-1	The Four Quantifier Negation (QN) Rules	
	Direct Statement	*Equivalent Statement*
All	$\forall x\,[Px]$	$\sim\exists x \sim[Px]$
Not all	$\sim\forall x\,[Px]$	$\exists x \sim[Px]$
Some	$\exists x\,[Px]$	$\sim\forall x \sim[Px]$
No	$\sim\exists x\,[Px]$	$\forall x \sim[Px]$

Note that in each case, the ~-operator that follows the quantifier doesn't change what's inside the brackets, but simply negates the entire contents.

Like the ten equivalence rules, **QN** works in both directions.

For example, you can start with this statement:

$\sim\exists x \sim[Gx \lor Hx]$

and end up with this one:

$\forall x\,[Gx \lor Hx]$

And also, just as you can with the ten equivalence rules, you can apply **QN** to part of a statement.

For example, you can change the following statement:

$\forall x\,[Mx \lor Lx]\ \&\ \exists x\,[Cx \lor Fx]$

to this:

$\sim\exists x \sim[Mx \lor Lx]\ \&\ \exists x\,[Cx \lor Fx]$

Using QN in proofs

With **QN** and the rules from SL, you can now prove the following argument:

$\forall x\,[Bx \to Cx] : \sim\exists x\,[Bx\ \&\ \sim Cx]$

1. $\forall x\,[Bx \to Cx]$ **P**

QN handles the big change involving the quantifier:

2. $\sim\exists x \sim[Bx \rightarrow Cx]$ 1 **QN**

The rest of the proof is just a couple of tweaks using the equivalence rules:

3. $\sim\exists x \sim[\sim Bx \vee Cx]$ 2 **Impl**

4. $\sim\exists x [Bx \,\&\, \sim Cx]$ 3 **DeM**

Because **QN** and the equivalence rules are all true in both directions, you could just as easily prove $\sim\exists x [Bx \,\&\, \sim Cx] : \forall x [Bx \rightarrow Cx]$ by reversing the steps. This reversal gives rigorous proof that these two statements mean the same thing — that is, they're semantically equivalent (for more information on semantic equivalence, see Chapter 6).

Table 17-2 encapsulates this information for the *all* and *not all* statements, listing four equivalent ways to write each type of statement.

Table 17-2	Four Equivalent Ways to Write *All* and *Not All* Statements	
	All	**Not All**
Direct statement	$\forall x [Bx \rightarrow Cx]$	$\sim\forall x [Bx \rightarrow Cx]$
Apply **QN**	$\sim\exists x \sim[Bx \rightarrow Cx]$	$\exists x \sim[Bx \rightarrow Cx]$
Applying **Impl**	$\sim\exists x \sim[\sim Bx \vee Cx]$	$\exists x \sim[\sim Bx \vee Cx]$
Applying **DeM**	$\sim\exists x [Bx \,\&\, \sim Cx]$	$\exists x [Bx \,\&\, \sim Cx]$

Table 17-3 lists the equivalent ways to write *some* and *no* statements.

Table 17-3	Four Equivalent Ways to Write *Some* and *No* Statements	
	Some	**No**
Direct statement	$\exists x [Bx \,\&\, Cx]$	$\sim\exists x [Bx \,\&\, Cx]$
Apply **QN**	$\sim\forall x \sim[Bx \,\&\, Cx]$	$\forall x \sim[Bx \,\&\, Cx]$
Applying **DeM**	$\sim\forall x [\sim Bx \vee \sim Cx]$	$\forall x [\sim Bx \vee \sim Cx]$
Applying **Impl**	$\sim\forall x [Bx \rightarrow \sim Cx]$	$\forall x [Bx \rightarrow \sim Cx]$

Exploring the Four Quantifier Rules

When you started using the rules for proofs in SL, you probably found that different rules were helpful for different things. For example, **Simp** and **DS** were good for breaking statements down, while **Conj** and **Add** were good for building them up.

A similar idea applies in QL. Two of the four quantifier rules are *instantiation rules,* as you can see in Table 17-4. These two rules allow you to remove the quantifier and brackets from a QL statement, breaking the statement down so that the SL rules can do their thing. The other two quantifier rules are *generalization rules.* These rules allow you to add quantifiers and brackets to build up the statements you need to complete the proof.

Table 17-4   The Four Quantifier Rules in QL and Their Limitations

Quantifier	Breaking Down	Building Up
∀	Universal Instantiation (**UI**)	Universal Generalization (**UG**)
	* Changes a bound variable to either a free variable or a constant.	* Changes a free variable (not a constant) to a bound variable.
		* This bound variable must not already appear in an earlier line in the proof justified by **EI** or an undischarged **AP**.
∃	Existential Instantiation (**EI**)	Existential Generalization (**EG**)
	* Changes a bound variable to a free variable (not a constant).	* Changes either a free variable or a constant to a bound variable.
	*This variable must not be free in an earlier line in the proof.	

Two of these rules — **UI** and **EG** — are relatively easy, so I go over these first. **UI** allows you to break down statements quantified with ∀, whereas **EG** allows you to build up statements using the ∃ quantifier.

After you master **UI** and **EG**, you'll be ready for **EI** and **UG**. These two rules are a little trickier because they include some limitations that **UI** and **EG** don't have. But, in principle, **EI** is just another break-down rule and **UG** is just another build-up rule. The following sections give you everything you need to know to figure out how to use these four rules.

Easy rule #1: Universal Instantiation (UI)

Universal Instantiation (**UI**) allows you to do the following:

- ✔ Free the bound variable in a ∀-statement by removing the quantifier and the brackets.

- ✔ Having freed this variable, uniformly change it wherever it appears to any individual constant or variable if you choose to do so.

For example, suppose you know that this statement is true:

All snakes are reptiles.

In QL, you can express this statement as

$\forall x\, [Sx \rightarrow Rx]$

Because you know a fact about *all* snakes, it's safe to say that a similar statement about a *specific* snake is also true:

If Binky is a snake, then Binky is a reptile.

UI allows you to make this leap formally:

$Sb \rightarrow Rb$

The result is a statement that looks a lot like an SL statement, which means you can use the 18 SL rules.

Warming up with a proof

As practice, this section gives you an example of a warm-up proof. Suppose you want to prove the validity of this argument:

Premises:

All elephants are gray.

Tiny is an elephant.

Conclusion:

Tiny is gray.

Translate this argument into QL as follows:

$$\forall x\,[Ex \rightarrow Gx],\ Et : Gt$$

As with an SL proof, begin by listing the premises:

1. $\forall x\,[Ex \rightarrow Gx]$ **P**
2. Et **P**

Now, "unpack" the first premise using **UI**:

3. $Et \rightarrow Gt$ 1 **UI**

Here, I first removed the $\forall x$ and the brackets. I also chose to change the variable x uniformly (that is, everywhere it appears in the statement) to the individual constant t. I chose t in this case because this is the constant that appears in line 2, which becomes useful in the next step.

Now you can complete the proof using an old, familiar rule from SL:

4. Gt 2, 3 **MP**

Valid and invalid uses of UI

UI gives you a lot of options for when you're scrounging for the next steps in a proof.

For example, suppose you're given the following premise:

$$\forall x\,[(Pa\ \&\ Qx) \rightarrow (Rx\ \&\ Sb)]$$

The simplest thing that **UI** allows you to do is to simply remove the quantifier and brackets and keep the same variable:

$$(Pa\ \&\ Qx) \rightarrow (Rx\ \&\ Sb)$$

UI also allows you to change the variable, as long as you make this change uniformly throughout the entire expression:

$(Pa \ \& \ Qy) \rightarrow (Ry \ \& \ Sb)$

Furthermore, with **UI** you change the variable to any constant you choose, even constants that already appear in the statement, as long as you make the change uniformly throughout the statement. For example:

$(Pa \ \& \ Qa) \rightarrow (Ra \ \& \ Sb)$

$(Pa \ \& \ Qb) \rightarrow (Rb \ \& \ Sb)$

$(Pa \ \& \ Qc) \rightarrow (Rc \ \& \ Sb)$

In each case, I've replaced the variable x uniformly with a constant — first a, then b, and finally c.

When using **UI**, you must replace the variable uniformly with a single choice of a constant or variable. And you must preserve constants from the original statement.

For example, here is an *invalid* use of **UI**:

$(Pa \ \& \ Qa) \rightarrow (Rb \ \& \ Sb)$ WRONG!

In this invalid statement, I've incorrectly changed one x to an a and the other to a b. These replacements are wrong because the changes must be uniform — you must replace every x with the same constant or the same variable throughout. Here's another invalid use of **UI**:

$(Px \ \& \ Qx) \rightarrow (Rx \ \& \ Sx)$ WRONG!

In this statement, I've incorrectly tampered with the constants a and b. These changes are wrong because UI only allows you to change the variable that's being quantified in the original statement — in this case, x. Other variables and constants in the original statement must remain the same.

Easy rule #2: Existential Generalization (EG)

Existential Generalization (**EG**) allows you to change either an individual constant or a free variable to a bound variable by adding brackets and quantifying it with ∃.

For example, suppose you know that this statement is true:

My car is white.

In QL, you can express this statement as

Wc

Because you have an example of a specific thing that is white, you know more generally that *something* is white, so it's safe to say

There exists an x such that x is white.

You can make this leap formally using **EG**:

$\exists x\,[Wx]$

Warming up with a proof

Just as **UI** allows you to break down QL statements at the beginning of a proof, **EG** allows you to build them back up at the end.

For example, suppose you want to prove that the following argument is valid:

$\forall x\,[Px \rightarrow Qx],\ \forall x\,[(Pb\ \&\ Qx) \rightarrow Rx],\ Pa\ \&\ Pb : \exists x\,[Rx]$

1.	$\forall x\,[Px \rightarrow Qx]$	**P**
2.	$\forall x\,[(Pb\ \&\ Qx) \rightarrow Rx]$	**P**
3.	$Pa\ \&\ Pb$	**P**

First, use **UI** to unpack the first two premises:

4.	$Pa \rightarrow Qa$	1 **UI**
5.	$(Pb\ \&\ Qa) \rightarrow Ra$	2 **UI**

Line 5 illustrates a point I discussed in the previous section: You can replace both instances of the variable x in statement 2 with the constant a, but the constant b must remain unchanged.

Next, use **Simp** to break down statement 3:

6.	Pa	3 **Simp**
7.	Pb	3 **Simp**

Now you can call forth all your old tricks from SL proofs:

8.	*Qa*	4, 6 **MP**
9.	*Pb & Qa*	7, 8 **Conj**
10.	*Ra*	5, 9 **MP**

You've now found one example of a constant (*a*) that has the property you're seeking (*R*), so you can complete the proof using **EG**:

11.	∃*x* [*Rx*]	10 **EG**

Valid and Invalid Uses of EG

EG gives you the same choices as **UI**, but with **EG** you're adding brackets and binding variables rather than removing brackets and freeing variables.

EG allows you to change any constant to a variable and then bind that variable using the quantifier ∃. As with **UI**, the changes you make must be uniform throughout. For example, suppose you have this premise:

(*Pa* & *Qb*) → (*Rb* ∨ *Sa*)

You can use **EG** to change either constant (*a* or *b*) to a variable and bind it with a ∃ quantifier:

∃*x* [(*Px* & *Qb*) → (*Rb* ∨ *Sx*)]

∃*x* [(*Pa* & *Qx*) → (*Rx* ∨ *Sa*)]

As with **UI**, when you use **EG**, the changes you make to constants must be uniform throughout. For example, here are three invalid uses of **EG**:

∃*x* [(*Px* & *Qx*) → (*Rx* ∨ *Sx*)]   WRONG!

In this invalid statement, I incorrectly changed both constants *a* and *b* to *x*.

∃*x* [(*Px* & *Qb*) → (*Rb* ∨ *Sa*)]   WRONG!

Here, I went the other direction and changed one *a* but not the other.

∃*x* [(*Pa* & *Qb*) → (*Rb* ∨ *Sa*)]   WRONG!

In this statement, I failed to change *any* constants to variables.

EG also allows you to make similar changes to statement forms. Statement forms won't be premises in an argument but they may arise from using **UI** and other quantifier rules. Suppose you have this statement form:

$$(Vx \ \& \ Ty) \lor (Cx \rightarrow Gy)$$

You can use **EG** to bind either variable:

$$\exists x \, [(Vx \ \& \ Ty) \lor (Cx \rightarrow Gy)]$$
$$\exists y \, [(Vx \ \& \ Ty) \lor (Cx \rightarrow Gy)]$$

You can also use **EG** to make a change of variable, as long as you make this change uniformly throughout:

$$\exists z \, [(Vx \ \& \ Tz) \lor (Cx \rightarrow Gz)]$$
$$\exists z \, [(Vz \ \& \ Ty) \lor (Cz \rightarrow Gy)]$$

Note that all four of these examples are still statement forms because in each case one variable is still free.

As with constants, when you use **EG** to bind or change variables, the changes need to be uniform throughout the statement form. Here are some invalid uses of **EG** applied to statement forms:

$$\exists x \, [(Vx \ \& \ Tx) \lor (Cx \rightarrow Gx)] \qquad \text{WRONG!}$$

In this invalid statement, I incorrectly used **EG** to bind both variables x and y using the variable x.

$$\exists z \, [(Vx \ \& \ Tz) \lor (Cx \rightarrow Gy)] \qquad \text{WRONG!}$$

Here, I changed one y to a z but failed to change the other y.

$$\exists z \, [(Vx \ \& \ Ty) \lor (Cx \rightarrow Gy)] \qquad \text{WRONG!}$$

In this statement, I bound the variable z but failed to change *any* variables to z.

Not-so-easy rule #1: Existential Instantiation (EI)

Existential Instantiation (**EI**) allows you to free the bound variable in an ∃-statement (or change it to a different variable) by removing the quantifier and the brackets — *provided* that this variable isn't already free in an earlier line of the proof.

You're right, this description is a mouthful. So, to keep it simple, I explain it bit by bit. And I start by explaining everything up to the word *provided*.

How EI is similar to UI

The first part of the description of **EI** is similar to **UI**, except that it works with ∃-statements (instead of ∀-statements). Suppose you have the following statement:

Something is green.

You can represent this statement in QL as follows:

∃x [Gx]

As with **UI**, **EI** allows you to free the bound variable, giving you this:

Gx

This statement form means "x is green," which is close to the meaning of the original statement. **EI** also allows you to change the variable:

Gy

In this case, the statement form means "y is green," is also close to what you started with because variables are just placeholders.

EI works with variables but not constants

Unlike **UI**, however, **EI** doesn't allow you to change the variable to a constant. For example, consider the following:

1.	∃x [Gx]	**P**
2.	Ga	**1 EI** WRONG!

The reason you can't use **EI** in this way is simple: Just because the original sentence said that *something* is green is no reason to assume that applesauce, armadillos, Alaska, Ava Gardner, or any other *specific* thing that a might represent is also green.

EI is more restrictive than **UI**. With **UI**, you start out knowing that *everything* has a given property, so you can conclude that any constant or variable also has it. But with **EI**, you start out knowing only that *something* has that property, so you can only conclude that a variable also has it.

An example using EI

To help you better understand how to use **EI**, consider the following argument:

$$\forall x \, [Cx \to Dx], \exists x \, [\sim Dx] : \sim \forall x \, [Cx]$$

1.	$\forall x \, [Cx \to Dx]$	**P**
2.	$\exists x \, [\sim Dx]$	**P**

First of all, use **UI** and **EI** to remove the quantifiers. But use **EI** *before* using **UI**. (This order has to do with stuff in the definition of **EI** that comes after the word *provided*, which I explain later.)

3.	$\sim Dx$	**2 EI**
4.	$Cx \to Dx$	**3 UI**

The next step is obvious, even if you're not quite sure yet how it helps:

5.	$\sim Cx$	**3, 4 MT**

Now, you know something about this variable x — namely, that x doesn't have the property C. So, you can use **EG** to make the more general statement "There exists an x such that x doesn't have the property C:

6.	$\exists x \, [\sim Cx]$	

To complete the proof, you can use **QN**. For clarity, in this case I take an extra step and explicitly use double negation (**DN**):

7.	$\sim \forall x \sim [\sim Cx]$	**6 QN**
8.	$\sim \forall x \, [Cx]$	**7 DN**

EI only allows you to free a variable that isn't free elsewhere

In the example in the previous section, I purposely used **EI** before I used **UI**. Now I explain why.

Note that the last part of the **EI** definition says "*provided* that this variable isn't already free in an earlier line of the proof." So, referring to the example in the previous section, if I had written $Cx \to Dx$ as line 3, I couldn't have used **EI** to write $\sim Dx$ in line 4.

This limitation may seem like a technicality, but trust me, it's important. To show you why, I pretend to prove an argument that's obviously false. Then I show you where the trouble comes from. Here's the argument:

Premises:

People exist.

Cats exist.

Conclusion:

Some people are cats.

The argument translated into QL is:

$\exists x\,[Px], \exists x\,[Cx] : \exists x\,[Px\ \&\ Cx]$

And here's the "proof":

1.	$\exists x\,[Px]$	**P**
2.	$\exists x\,[Cx]$	**P**
3.	Px	1 **EI**
4.	Cx	2 **EI** WRONG!
5.	$Px\ \&\ Cx$	3, 4 **Conj**
6.	$\exists x\,[Px\ \&\ Cx]$	5 **EG**

There must be a catch. Otherwise, you'd find some cat people wandering around with the rest of the human race. (I wonder if they get hairballs.) So, where does the problem come from?

Line 3 is fine: I use **EI** to free the variable x. But then in line 4, I try to use **EI** to free the variable x again. This move is a big no-no, and it leads to the mess that results. Because I've *already* freed the variable x in line 3, I'm not allowed to use this variable with **EI** in line 4.

I could have written Py or Pz in line 4, but then the rest of the argument would have fallen apart. But, this is a good thing, because bad arguments are *supposed* to fall apart.

Valid and Invalid Uses of EI

In this section, I clarify exactly how and when you can use **EI**.

For example, suppose you're given these two premises:

1. $\exists x [(Rx \& Fa) \rightarrow (Hx \& Fb)]$ **P**
2. $\exists y [Ny]$ **P**

As with **UI**, **EI** allows you to free a variable, with the option of changing it along the way. Here are three valid next steps:

3. $(Rx \& Fa) \rightarrow (Hx \& Fb)$ 1 **EI**
4. $(Ry \& Fa) \rightarrow (Hy \& Fb)$ 1 **EI**
5. Nz 2 **EI**

In line 3, I used **EI** to free the variable x. This is the most common use of **EI**. In line 4, I changed the variable from x to y and then freed it using **EI**. This is also a valid option. And then in line 5, I wanted **EI** to free the variable from line 2. But because I had *already* freed x and y in previous lines, I needed to pick a new variable z.

As with **UI**, if you change the variable using **EI**, you must make uniform changes throughout. For example, check out this invalid statement:

6. $(Rv \& Fa) \rightarrow (Hw \& Fb)$ 1 **EI**   WRONG!

In line 6, I incorrectly changed one x to v and the other to w. This change is just as wrong with **EI** as it would be with **UI**.

But, **EI** limits your options in ways that **UI** doesn't. For example, take a look at this invalid statement:

7. $(Ra \& Fa) \rightarrow (Ha \& Fb)$ 1 **EI**   WRONG!

In line 7, I uniformly changed the variable x to the individual constant a. Not allowed! With **EI**, variables must remain variables. Take a look at the last invalid statement:

8. Nx 2 **EI**   WRONG!

In line 8, I changed the variable y to x. But the variable x already appears as a free variable in line 3, so you're not allowed to use it a second time with **EI**. This type of error is the most common one you're likely to make with **EI**, so watch out for it.

Not-so-easy rule #2: Universal Generalization (UG)

Universal Generalization (**UG**) allows you to change a free variable to a bound variable by adding brackets and quantifying it with ∀ — *provided* that this variable isn't free in an earlier line of the proof that's justified either by **EI** or by an undischarged AP (flip to Chapter 11 for more on discharging an **AP**).

As with **EI**, **UG** is fairly straightforward up to the word *provided*. To make it as easy as possible, I explain it piece by piece.

How UG is similar to EG

For the most part, **UG** is similar to **EG**. The biggest difference is that **EG** works with ∃ statements (instead of ∀-statements, which **UG** uses). Suppose you're given the following definition:

Let Nx = x is nice

Suppose further at some point in a proof, you arrive at the following statement form:

Nx

Under the proper circumstances, **UG** allows you to derive the more general statement "For all x, x is nice," or, more simply put, "Everything is nice":

$\forall x \, [Nx]$

The variable you choose in this case doesn't really matter, so you can also use **UG** to write:

$\forall y \, [Ny]$

UG works with variables but not constants

Unlike **EG**, **UG** doesn't allow you to change a constant to a variable. For example, consider the following:

1.	Nc	**P**	
2.	$\forall x \, [Nx]$	1 UG	WRONG!

The reason this statement is wrong is simple: Just because the original sentence said that whatever *c* stands for (cars, cows, Cadillacs, or Candice Bergen) is nice, you can't jump to the conclusion that *everything* is also nice.

UG is more restrictive than **EG**. With **EG**, you want to show that *something* has a given property by finding one specific thing — either a constant or a variable — that has that property. With **UG**, however, you want to show that *everything* has that property by finding a variable that has it.

An example using UG

As with **EG**, **UG** is most often used at the end of a proof to build a statement form that you've pieced together into a statement by quantifying it with ∀. Consider this argument:

$$\forall x\,[Bx \rightarrow Cx],\ \forall y\,[\sim Cy \vee Dy] : \forall x\,[Bx \rightarrow Dx]$$

1.	$\forall x\,[Bx \rightarrow Cx]$	**P**
2.	$\forall y\,[\sim Cy \vee Dy]$	**P**

The first step is to break open the two premises using **UI**:

3.	$Bx \rightarrow Cx$	1 **UI**
4.	$\sim Cx \vee Dx$	2 **UI**

Note that in line 4, **UI** provides the freedom to change the variable from *y* to *x* without worrying that *x* is already free in line 3. And once the variables are all the same, you can move forward as if you were writing an SL proof:

5.	$Cx \rightarrow Dx$	4 **Impl**
6.	$Bx \rightarrow Dx$	3, 5 **HS**

Now, everything is set up for completing the proof with **UG**:

7.	$\forall x\,[Bx \rightarrow Dx]$	6 **UG**

UG only allows you to bind a variable that isn't free on a line justified by either EI or an undischarged AP

So, here's my confession: I think the rule **UG** is just about the orneriest concept in this whole book. Okay, now I've said it. And what makes **UG** a problem is this rule about how it's *not* supposed to be used.

But, look on the bright side: when you've mastered **UG**, you can rest assured that nothing worse is in store for you.

First off, imagine that you start a club that has one member: you. Even though you start off small, you have big plans for the club: Eventually, you want everyone in the whole world to join. Your sense of logic is somewhat twisted, so you make this argument:

Premise:

A member of my club exists.

Conclusion:

Everything is a member of my club.

Then, you translate this argument into QL as follows:

$\exists x\,[Mx] : \forall x\,[Mx]$

This plan sounds a little fishy, but then you "prove" your argument like this:

1.	$\exists x\,[Mx]$	**P**	
2.	Mx	**1 EI**	
3.	$\forall x\,[Mx]$	**2 UG**	WRONG!

Somehow, in two easy steps, you increased your club membership from a single member to everything in the entire universe. Just imagine what UNICEF could do with this argument!

Here's the problem: The variable x is free on line 2, which is justified by **EI**, so you can't use **UG** to bind x in line 3. This scenario is exactly what the definition of **UG** after the word *provided* is getting at.

As another example, imagine that you're *not* a billionaire. (If this is difficult to imagine, repeat after me: "I am not a billionaire.") From this premise, you're going to try to prove that Donald Trump isn't a billionaire. Here's your argument:

Premise:

I am not a billionaire.

Conclusion:

Donald Trump isn't a billionaire.

The translation of this argument into QL is as follows:

~Bi : ~Bt

And, now, here's the "proof":

1.	~Bi	P
2.	Bx	AP

Here, I use an advanced indirect proof strategy that I discuss more in Chapter 12: I make up an assumed premise (**AP**) and then work to prove a contradiction. If I'm successful, I have proved the *negation* of the **AP**. In general, this is a perfectly valid strategy, but the next step contains a fatal error:

3.	∀x [Bx]	1 UG WRONG!
4.	Bi	2 UI
5.	Bi & ~Bi	1, 4 **Conj**
6.	~Bx	2–5 **IP**

Having discharged my **AP**, the rest of the proof looks simple:

7.	∀x [~Bx]	6 UG
8.	~Bt	7 UI

The problem this time is that the variable x is free in line 2, which is justified by an *undischarged* **AP**, which means that you can't use **UG** to bind x.

Notice, however, that in line 7, the proof includes a perfectly valid usage of **UG** to bind x. The only difference here is that by this point in the proof, the **AP** has been discharged.

The moral of the story is that whenever you want to use **UG**, you need to check first to make sure that the variable you're trying to bind is not free in either of the following:

 ✔ A line that's justified by **EI**

 ✔ A line that's justified by an **AP** that has *not* been discharged

Valid and Invalid Uses of UG

Here is a summary of when you can and can't use **UG**. Suppose you have the following proof in progress (this takes a little setting up):

1.	$\forall x\,[Tx]$	**P**
2.	$\exists y\,[Gy]$	**P**
3.	Sa	**P**
4.	Tx	1 **UI**
5.	Gy	2 **EI**
6.	$Tx \,\&\, Sa$	3, 4 **Conj**
7.	$(Tx \,\&\, Sa) \,\&\, Gy$	5, 6 **Conj**
8.	Hz	**AP**

Now, you can use **UG** to bind the variable x in line 7:

9.	$\forall x\,[(Tx \,\&\, Sa) \,\&\, Gy]$	7 **UG**

But, you *can't* use **UG** to change the constant a in line 7 to a variable and bind it:

10.	$\forall w\,[(Tx \,\&\, Sa) \,\&\, Gy]$	7 **UG**	WRONG!

You also *can't* use **UG** to bind the variable y in line 7, because this variable appears free on line 5, which is justified by **EI**:

11.	$\forall y\,[(Tx \,\&\, Sa) \,\&\, Gy]$	7 **UG**	WRONG!

Furthermore, you *can't* use **UG** to bind the variable z in line 8, because this variable is justified by an **AP** that is still undischarged:

12.	$\forall z\,[Hz]$	8 **UG**	WRONG!

Chapter 18

Good Relations and Positive Identities

. . . .

In This Chapter

▶ Understanding relational expressions

▶ Discovering identities

▶ Writing proofs using relations and identities

. . . .

*I*n sentential logic (SL), you represent a basic statement with just one letter — a constant. For example, you can represent the statement "Barb is a tree surgeon," as

B (SL TRANSLATION)

In quantifier logic (QL), you represent the same sentence with a two-letter combination of a property constant and an individual constant, for example:

Tb

This representation works great. But, suppose you want to translate the statement "Barb hired Marty" into QL. Now that more than one person is in the picture, you need a way to express their relationship.

Or suppose you want to translate the statement "Barb is the best tree surgeon in the county." In this case, you aren't describing a property that applies to Barb, but rather her identity as the *single* best tree surgeon in the county.

Fortunately, QL allows you to express these more sophisticated ideas easily. In this chapter, I discuss both relations and identities in QL. *Relations* allow you to express statements that have more than one key player, and *identities* help you handle situations where the key player is identified with an alternative but uniquely-fitting description. Finally, I show you how these new concepts fit into the whole picture of QL.

Relating to Relations

Up until now, you've been using *monadic expressions:* Statements and statement forms that have only one individual constant or variable. For example, the statement

Na

has only one individual constant: *a*. Similarly, the statement form

Nx

has only one variable: *x*.

In this section, I help you expand your notion of expressions to include those with more than one individual constant or variable.

Defining and using relations

A *relational expression* has *more than one* individual constant or variable.

Defining a relational expression is quite easy after you know the basics of defining a monadic expression (see Chapter 15). For example, you can define a monadic expression as follows:

Let *Nx* = *x* is nosy

Then, you might translate the sentence "Arkady is nosy" as

Na

But, suppose you have the statement:

Arkady is nosier than Boris.

In this case, you can begin by defining the following relational expression:

Let *Nxy* = *x* is nosier than *y*

Then, you can use this expression to translate the statement as follows:

Nab

Similarly, you can translate the statement

> Boris is nosier than Arkady

as

> *Nba*

The order of the individual constants or variables in a relational expression is crucial — don't mix them up!

Connecting relational expressions

You can use the five SL operators (~, &, ∨, →, and ↔) to connect relational statements and statement forms just as with monadic expressions. (For more on distinguishing statements from statement forms, flip to Chapter 15.)

After you've translated a statement from English into a relational expression, it becomes an indivisible unit just like a monadic expression. From then on, the same rules apply for using the five operators.

For example, suppose you want to say:

> Kate is taller than Christopher, but Christopher is taller than Paula.

First, define the constants:

> Let *k* = Kate
>
> Let *c* = Christopher
>
> Let *p* = Paula

Next, define the appropriate relational expression:

> Let *Txy* = *x* is taller than *y*

Now, you can translate the statement

> *Tkc* & *Tcp*

Similarly, you can translate the statement

> If Kate is taller than Christopher, then she is also taller than Paula.

as

> *Tkc* → *Tkp*

Making use of quantifiers with relations

You can use the two QL quantifiers (∀ and ∃) to connect relational statements and statement forms just as with monadic expressions.

Using quantifiers with relations isn't substantially different from using them with monadic expressions.

For example, suppose you want to translate the following English statement into QL:

> Everybody likes chocolate.

First define the relational expression:

> Let *Lxy* = *x* likes *y*

Then, define the individual constant:

> Let *c* = chocolate

Now, you're ready to translate the statement as follows:

> ∀*x* [*Lxc*]

Similarly, suppose you have the following expression:

> Someone introduced Dolly to Jack.

This time, you need to define a relational expression that takes three individual variables:

> Let *Ixyz* = *x* introduced *y* to *z*

Using the initials for Dolly and Jack as individual constants, and binding the variable *x* with the quantifier ∃, you can translate the statement:

> ∃*x* [*Ixdj*]

You can also translate the statement

> Jack introduced Dolly to everyone.

as

> ∀*x* [*Ijdx*]

Notice that you can still use the variable *x* here even though the definition for the relational expression uses *y* in this position. The choice of variable isn't important here as long you've quantified it properly.

Working with multiple quantifiers

Because relations have more than one variable, they open up the possibility of statements with more than one quantifier. For example, you can translate the statement

> Everybody introduced someone to Jack.

as

> ∀x ∃y [*Ixyj*]

By breaking down this statement, you realize that it literally states: "For all *x*, there exists a *y* such that *x* introduced *y* to Jack."

Similarly, if you want to translate the following statement into QL:

> Jack introduced someone to everyone.

You can use the following statement:

> ∀x ∃y [*Ijyx*]

This statement means "For all *x*, there exists a *y* such that Jack introduced *y* to *x*."

Be careful with the order of quantifiers. Changing the order of quantifiers from ∀x ∃y to ∃y ∀x changes the meaning of statement even if the contents of the brackets remains the same.

Here is the distinction between statements starting with ∀x ∃y and those starting with ∃y ∀x

✔ Generally speaking, ∀x ∃y means "For all *x*, there exists a *y* such that . . .", which means that *y may be different* for two different *x*'s.

✔ On the other hand, ∃y ∀x means "There exists a *y* such that for all *x* . . .", which means that the *y is the same* for every *x*.

To make this distinction clear, I define a new property constant:

Let *Mxy* = *x* is married to *y*

Now, suppose I want to express the idea that all people are married. I can translate the statement as

∀*x* ∃*y* [*Mxy*]

This statement literally means "For all *x*, there exists a *y* such that *x* is married to *y*."

But suppose I reverse the quantifiers:

∃*y* ∀*x* [*Mxy*]

Now, it means "There exists a *y* such that for all *x*, *x* is married to *y*." In this case, I'm saying that all people are married to *the same person*. This is definitely not what I had in mind.

Multiple quantifiers are especially useful to express mathematical ideas. For example, suppose you want to express the idea that the counting numbers (1, 2, 3 . . .) are infinite. You can express this idea by defining a relational expression:

Let *Gxy* = *x* is greater than *y*

Now, you can express the infinity of all numbers by stating that for every number, there exists a number that's greater:

∀*x* ∃*y* [*Gyx*]

In other words, "For all *x*, there exists a *y* such that *y* is greater than *x*."

Writing proofs with relations

When you're writing proofs, relational expressions work pretty much the same as monadic expressions. They stand as discrete and inseparable chunks that you can manipulate using the 18 rules of inference for SL (see Chapters 9 and 10).

Quantifier negation (**QN**) and the four quantifier rules also work the same with relational expressions as they do with monadic expressions (I cover these fabulous results in logic in Chapter 17). The main difference is that now you may need to write proofs that include statements with multiple quantifiers.

Using QN with multiple quantifiers

QN works the same with multiple quantifiers as it does with single quantifiers. (Flip to Chapter 17 for a refresher on using **QN**.) Just make sure you are clear which quantifier you are changing.

For example, here's a new declaration:

Let $Pxy = x$ gave a present to y

Suppose you want to prove the following argument:

Premise:

Everyone gave at least one person a present.

Conclusion:

Nobody didn't give a present to anybody.

You can write this argument in QL as follows:

$\forall x \, \exists y \, [Pxy] : {\sim}\exists x \, \forall y \, {\sim}[Pxy]$

1.	$\forall x \, \exists y \, [Pxy]$	**P**
2.	${\sim}\underline{\exists x} \, {\sim}\exists y \, [Pxy]$	1 QN

In line 2, I underlined the first quantifier because it's the focus of the change from line 1. Notice that I placed a ~-operator before this quantifier, changed the quantifier from $\forall x$ to $\exists x$, and placed another ~-operator after this quantifier.

3.	${\sim}\exists x \, {\sim}{\sim}\underline{\forall y} \, {\sim}[Pxy]$	2 QN
4.	${\sim}\exists x \, \underline{\forall y} \, {\sim}[Pxy]$	3 DN

In line 3, the second quantifier is now the focus of the change from line 2. Again, I've added ~-operators before and after this quantifier and changed it from $\exists y$ to $\forall y$. To keep the proof crystal clear, I've taken an additional line to apply the double negation rule (**DN**).

Using the four quantifier rules with multiple quantifiers

When using the four quantifier rules — UI, EG, EI, and UG (see Chapter 17) — with multiple quantifiers, an important restriction comes into play: You can only remove the first (leftmost) quantifier or add a new first quantifier.

In practice, this restriction means that you have to do the following:

✔ Break statements down from the outside in.

✔ Build statements up from the inside out.

For example, take a look at this proof:

$\forall x\, \forall y\, [Jxy \rightarrow Kyx]$, $\exists x\, {\sim}\exists y\, [Kyx] : \exists x\, \forall y\, [{\sim}Jxy]$

| 1. | $\forall x\, \forall y\, [Jxy \rightarrow Kyx]$ | P |
| 2. | $\exists x\, {\sim}\exists y\, [Kyx]$ | P |

First, a quick switch of quantifiers in the second premise to move the ~-operator to the right and get it out of the way:

| 3. | $\exists x\, \forall y\, {\sim}[Kyx]$ | 2 QN |

But, this premise still has a $\exists$-quantifier, so I have to use existential instantiation (EI). I need to get this step out of the way as soon as possible before x shows up free in the proof (see Chapter 17 for more on EI). So, now I go about breaking it down from the outside in:

| 4. | $\forall y\, {\sim}[Kyx]$ | 3 EI |
| 5. | ${\sim}Kyx$ | 4 UI |

Now I'm at the first premise, again breaking down from the outside in:

| 6. | $\forall y\, [Jxy \rightarrow Kyx]$ | 1 UI |
| 7. | $Jxy \rightarrow Kyx$ | 6 UI |

The next step is pretty clear:

| 8. | ${\sim}Jxy$ | 5, 7 MT |

Time to build the conclusion from the inside out:

| 9. | $\forall y\, [{\sim}Jxy]$ | 8 UG |

Regarding self-referential statements

A few fine points to proofs with relations are beyond the scope of this book. Many of these issues arise with the issue of *self-referential statements*, which are relational expressions with a repeated individual constant.

For example, when given the statement "Jack loves everybody" — $\forall x [Ljx]$ — you can use **UI**

to infer the statement "Jack loves himself" — *Ljj*.

Self-referential statements are necessary to make QL fully expressive, but you need to keep a close eye on them in proofs.

This use of universal generalization (**UG**) is valid because in line 4, the variable *y* is still bound. Finally you use existential generalization (**EG**) to finish off the proof:

10. $\exists x \, \forall x \, [\sim Jxy]$ 9 **EG**

Identifying with Identities

Look at the following two statements:

George Washington was a U.S. president.

George Washington was the first U.S. president.

You can translate the first statement into QL easily as

Pg

At first glance, you might think that you could handle the second statement with the same approach. But, you can't because these two statements, even though they look similar, are actually quite different.

The first statement describes Washington in terms of a *property* that's possibly shared by others (for example, by Abraham Lincoln). That's why you can easily use a *property constant* to translate this statement into QL (see Chapter 15 for more on property constants).

The second statement, however, tells us about Washington in terms of his *identity* as the one and only first U.S. president. To translate this statement into QL, you need something new — and this something new is a way to express identity. I discuss the ins and outs of identities in the following sections.

Indirect discourse and identity

There are intriguing exceptions to the idea that *identity* means you can substitute freely from one statement to another without changing the meaning. For example, look at these two statements:

> Sally believed that Abraham Lincoln was the first U.S. president.

> Sally believed that Abraham Lincoln was George Washington.

Clearly, these statements don't mean the same thing, and the first statement can be true and the second false.

For this reason, QL doesn't work with statements that contain *indirect discourse*, such as

Elton *hoped* that Mary had paid the gas bill.

The boss *insisted* that everybody arrive on time.

We *know* that Clarissa is lying.

It is *necessary* that the sun rise every morning.

See Chapter 21 for a discussion of non-classical systems of logic that attempt to account for statements of these kinds.

But remember that, as far as QL is concerned, once you define an identity between two constants, you are saying it's OK to substitute one constant for the other.

Understanding identities

An *identity* tells you that two different individual constants refer to the same thing, meaning that they're interchangeable in QL.

What does it mean to say that George Washington was *the* first U.S. president? Essentially, you're saying that anywhere you talk about *George Washington*, you can substitute the words *the first U.S. president,* and vice versa.

In QL, you can state an identity formally as follows (assuming *g* stands for *George Washington* and *f* stands for *the first U.S. president*):

$$g = f$$

You can also use an identity with a quantifier:

$$\exists x\ [x = f]$$

This statement translates to: "There exists an *x* such that *x* is the first U.S. president," or, more simply, "The first U.S. president exists."

Writing proofs with identities

QL contains two rules just for handling identities. They're so easy to understand and use that I give you just one example of each, and you'll be all set.

Rule of Identity (ID)

ID just provides formally for the substitution of one constant for another in a proof after the identity of these constants has been shown.

After you establish an identity of the form $x = y$ in a proof, **ID** allows you to rewrite any line of that proof by substituting x for y (or y for x).

Here's an argument that requires an identity:

Premises:

> Every person deserves respect.
>
> Steve Allen is a person.
>
> Steve Allen was the original host of *The Tonight Show.*

Conclusion:

> The original host of *The Tonight Show* deserves respect.

Here's the same argument translated into QL:

$$\forall x\,[Px \rightarrow Dx],\, Ps,\, s = o : Do$$

The proof of this argument is straightforward:

1.	$\forall x\,[Px \rightarrow Dx]$	**P**
2.	Ps	**P**
3.	$s = o$	**P**
4.	$Ps \rightarrow Ds$	1 **UI**
5.	Ds	2, 4 **MP**
6.	Do	5 **ID**

I used **ID** on the last step to substitute o for s. In most cases, whenever you need to use **ID**, you'll be able to see the opportunity from miles away.

Identity Reflexivity (IR)

IR is even easier to understand than **ID**. It just tells you that it's okay to assume that everything is identical to itself, no matter what you're proving.

IR allows you to insert the statement $\forall x\,[x = x]$ at any point in a proof.

This statement $\forall x\,[x = x]$ translates into English as "For all x, x is identical to x," or, more simply, "Everything is identical to itself."

IR is one of those rules you're almost never going to need unless the teacher sets up a proof on an exam that requires it (which is always possible). For example, take a look at the following proof:

$\forall x\,[((x = m) \lor (x = n)) \to Tx] : Tm \;\&\; Tn$

1.	$\forall x\,[((x = m) \lor (x = n)) \to Tx]$	**P**
2.	$((m = m) \lor (m = n)) \to Tm$	1 **UI**
3.	$((n = m) \lor (n = n)) \to Tn$	1 **UI**

I used **UI** to unpack the premise in two different ways: In line 2, changing the variable x to the constant m, and in line 3 changing x to n. Now, you can pull out **IR**:

4.	$\forall x\,[x = x]$	**IR**

IR is like an extra premise, so you don't need to reference any line numbers when you use it. After using **IR**, you can use **UI** to get the identity statements you need:

5.	$m = m$	4 **UI**
6.	$n = n$	4 **UI**

This time, I used **UI** to unpack line 4 in two different ways, again changing x first to m and then to n.

Now everything is in place to finish the proof using only rules of inference from SL:

7.	$(m = m) \lor (m = n)$	5 **Add**
8.	$(n = m) \lor (n = n)$	6 **Add**
9.	Tm	2, 7 **MP**
10.	Tn	3, 8 **MP**
11.	$Tm \;\&\; Tn$	9, 10 **Conj**

Chapter 19

Planting a Quantity of Trees

● ●

In This Chapter

▶ Extending the truth tree method to QL statements

▶ Understanding non-terminating truth trees

● ●

*I*n Chapter 8, I show you how to use truth trees in sentential logic (SL) for a variety of different purposes. In this chapter, you see how this method can be extended to quantifier logic (QL). As in SL, truth trees in QL are generally simpler than proofs. You don't need to have a brainstorm to make them work — just plug and chug. I warn you right up front that (unfortunately) QL truth trees have limitations.

In this chapter, I show you how to make the most of the truth tree method for solving problems in QL. I also show you an important drawback — the non-terminating tree.

Applying Your Truth Tree Knowledge to QL

Everything you know about building truth trees in SL also applies to QL trees. In this section, I give you a QL example to show you how it's done. If at anytime you come across something unfamiliar and you get stuck, flip to Chapter 8 for a refresher.

Using the decomposition rules from SL

Suppose you want to test whether the following set of three statements is consistent:

$Ma \lor \sim Tb$

$\sim Ma \mathbin{\&} Lc$

$\sim Tb$

As with trees in SL, the first step in deciding whether statements are consistent or inconsistent is to build the trunk of the tree using these three statements:

$$Ma \lor \sim Tb$$
$$\sim Ma \,\& \, Lc$$
$$\boxed{\sim Tb}$$

Notice that I circled the third statement because it's already in a form that can't be decomposed further.

Now, I can build the tree using the decomposition rules from SL. I start with the second statement because it's a single-branching statement:

$$Ma \lor \sim Tb$$
$$\sim Ma \,\& \, Lc \checkmark$$
$$\boxed{\sim Tb}$$
$$\boxed{\sim Ma}$$
$$\boxed{Lc}$$

Having decomposed the statement $\sim Ma \,\& \, Lc$, I checked this statement to show that I'm done with it. And again, I circled the resulting statements that are fully decomposed. The only statement left that isn't checked or circled is the first statement:

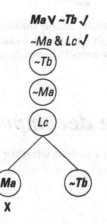

$$Ma \lor \sim Tb \checkmark$$
$$\sim Ma \,\& \, Lc \checkmark$$
$$\boxed{\sim Tb}$$
$$\boxed{\sim Ma}$$
$$\boxed{Lc}$$
$$Ma \qquad \sim Tb$$
$$X$$

Now, every statement is either checked or circled, so the tree is finished.

Notice that I also closed off the branch that ends with *Ma* by placing an X below it. As with SL trees, the reason for closing off this branch is simple: To travel all the way from the beginning of the trunk to the end of this branch, you need to pass through both *~Ma* and *Ma*. Because a statement and its negation can't both be true, the branch is closed off. And this is the only reason to close off a branch.

But the branch that ends with *~Tb* has no such contradiction, so I left it open. As with SL trees, because this truth tree has at least one open branch, the set of three statements is considered *consistent*.

Adding UI, EI, and QN

For statements with quantifiers, you need to add in some of the QL quantifier rules from Chapter 17. For breaking down statements in trees, use universal instantiation (**UI**) and existential instantiation (**EI**). And for removing the ~-operator from a quantifier, use quantifier negation (**QN**).

Here's an argument with one premise, which I test for validity using a truth tree:

$\sim\exists x\ [Gx\ \&\ \sim Nx] : \forall x\ [Gx \rightarrow Nx]$

As always, the first step is to build the trunk of the tree using the premise and the negation of the conclusion:

$$\sim\exists x\ [Gx\ \&\ \sim Nx]$$
$$\sim\forall x\ [Gx \rightarrow Nx]$$

Because both statements have negated quantifiers, use **QN** to move the ~-operators. (Note that you don't have to use line justifications such as **QN**, **UI**, or **EI** with a tree as you do with a proof.) I save space here by taking two steps at once:

$$\sim\exists x\ [Gx\ \&\ \sim Nx]\ \checkmark$$
$$\sim\forall x\ [Gx \rightarrow Nx]\ \checkmark$$
$$\forall x\ \sim[Gx\ \&\ \sim Nx]$$
$$\exists x\ \sim[Gx \rightarrow Nx]$$

Now you're ready to use **UI** and **EI**. However, remember the limitation on **EI**: Just as in QL proofs, you can only use **EI** to free a variable that isn't already free. So, in this example, you have to get **EI** out of the way before using **UI**:

$$\sim\exists x\,[Gx\,\&\,\sim Nx]\,\checkmark$$
$$\sim\forall x\,[Gx\rightarrow Nx]\,\checkmark$$
$$\forall x\,\sim[Gx\,\&\,\sim Nx]$$
$$\exists x\,\sim[Gx\rightarrow Nx]\,\checkmark$$
$$\sim[Gx\rightarrow Nx]$$
$$\sim[Gx\,\&\,\sim Nx]$$

When you use **UI**, don't check the statement that you just decomposed. I tell you why later in the chapter. But, for now, just remember the rule: When you use **EI**, check the decomposed statement as usual, but don't check it when you use **UI**.

At this point, you can start decomposing statements using the rules from SL truth trees. When possible, start with a single-branching statement:

$$\sim\exists x\,[Gx\,\&\,\sim Nx]\,\checkmark$$
$$\sim\forall x\,[Gx\rightarrow Nx]\,\checkmark$$
$$\forall x\,\sim[Gx\,\&\,\sim Nx]$$
$$\exists x\,\sim[Gx\rightarrow Nx]\,\checkmark$$
$$\sim[Gx\rightarrow Nx]\,\checkmark$$
$$\sim[Gx\,\&\,\sim Nx]$$
$$(Gx)$$
$$(\sim Nx)$$

Finally, decompose the double-branching statement:

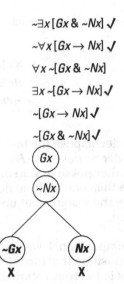

$\sim\exists x [Gx \& \sim Nx]$ ✓

$\sim\forall x [Gx \rightarrow Nx]$ ✓

$\forall x \sim[Gx \& \sim Nx]$

$\exists x \sim[Gx \rightarrow Nx]$ ✓

$\sim[Gx \rightarrow Nx]$ ✓

$\sim[Gx \& \sim Nx]$ ✓

Gx

$\sim Nx$

$\sim Gx$ Nx

X   X

Both branches lead to contradictions, so they both get closed off with X's. Because every branch has been closed off, the tree is finished and the argument is considered valid.

Using UI more than once

In the example from the previous section, I told you that when you use **UI** to decompose a statement, you must *not* check it as complete. I promised an explanation, and here it comes.

When you decompose a ∀-statement using **UI**, you have an unlimited number of constants you can choose for the decomposition. So, leave the statement unchecked because you may need to use it again.

For example, consider this argument:

$\forall x [Hx \rightarrow Jx]$, $Ha \& Hb : Ja \& Jb$

Set up the trunk of the tree as usual:

$$\forall x [Hx \rightarrow Jx]$$
$$Ha \;\&\; Hb$$
$$\sim(Ja \;\&\; Jb)$$

I know I need to decompose the first premise using **UI**, but I have many choices. I could decompose it to $Hx \rightarrow Jx$, or $Ha \rightarrow Ja$, or $Hb \rightarrow Jb$, or an infinite number of other possible statements. As you will see in this example, I may need more than one of these decompositions to complete the tree, so I have to leave the ∀-statement unchecked so that I am free to decompose it again as needed.

For my first decomposition, I want to go in a direction that looks useful, so I choose a decomposition that includes one of the constants that appears in other statements in the tree. I start with the constant *a*:

$$\forall x [Hx \rightarrow Jx]$$
$$Ha \;\&\; Hb$$
$$\sim(Ja \;\&\; Jb)$$
$$Ha \rightarrow Ja$$

I left the statement that I just decomposed *unchecked*, and now I move on to the next step:

$$\forall x [Hx \rightarrow Jx]$$
$$Ha \;\&\; Hb \;\checkmark$$
$$\sim(Ja \;\&\; Jb)$$
$$Ha \rightarrow Ja$$

$$\left(Ha\right)$$
$$\left(Hb\right)$$

This time, I checked the statement as usual. Now I have to double-branch:

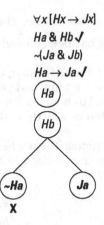

$\forall x [Hx \rightarrow Jx]$
$Ha \mathbin{\&} Hb \checkmark$
$\sim(Ja \mathbin{\&} Jb)$
$Ha \rightarrow Ja \checkmark$
Ha
Hb
$\sim Ha$ Ja
X

Fortunately, one of the branches gets closed off, and I get started on the next step:

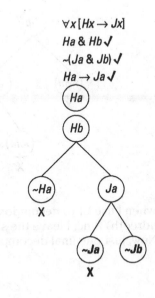

$\forall x [Hx \rightarrow Jx]$
$Ha \mathbin{\&} Hb \checkmark$
$\sim(Ja \mathbin{\&} Jb) \checkmark$
$Ha \rightarrow Ja \checkmark$
Ha
Hb
$\sim Ha$ Ja
X
$\sim Ja$ $\sim Jb$
X

You may be tempted at this point to think that the tree is finished. But remember that the tree is only finished when:

- Every item has been either checked or circled.
- Every branch has been closed off.

If these rules don't sound familiar, check out Chapter 8. Here's the punch line: The reason the tree isn't finished yet is because when I used **UI** to decompose the first statement, I didn't check it off as complete. So, now, I decompose this statement in a different way that allows me to complete the tree.

In this case, I decompose the unchecked ∀-statement using the other constant that appears in the tree — namely, the constant *b*:

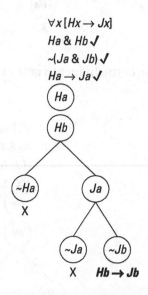

Note that even when I use **UI** to decompose the first statement for the second (or third, or hundredth) time, I leave the statement *unchecked* in case I need to use it again. Now, just one final decomposition and I'm done:

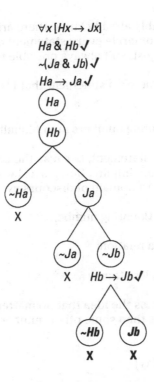

$\forall x [Hx \rightarrow Jx]$

$Ha \ \& \ Hb \checkmark$

$\sim(Ja \ \& \ Jb) \checkmark$

$Ha \rightarrow Ja \checkmark$

With this last decomposition, every branch is closed off, which means I'm done. Because all of the branches are closed off, you know the argument is valid.

Non-Terminating Trees

SL trees are my favorite logical tool because they don't require you to be clever. You just have to follow the right steps to the end and you get the correct answer every time.

Unfortunately, QL trees tend to be a bit more unruly. In some cases, a tree just grows and grows and grows — and never stops growing. This type of tree is called a *non-terminating tree* (or an *infinite tree*), and it makes things interesting, if not somewhat frustrating.

As you probably already know, there are only two ways to finish a tree: Either check or circle every statement *or* close off every branch with an X. But, in QL, it just isn't always possible to complete a tree.

To illustrate, here's a statement that I translate into QL and then test using a truth tree:

The counting numbers aren't infinite.

This is a *false* statement, because the counting numbers (1, 2, 3. . .) go on forever, so they are infinite. But, it's a statement that you can express in QL. First, I declare a domain of discourse:

Domain: Counting numbers

Now, I define a relation:

Lxy = *x* is less than *y*

I want to express the idea that no matter what number *x* I choose, that number is less than some other number *y*, and then negate it. Here's how I write it:

~∀*x* ∃*y* [*Lxy*]

This translates literally as "It isn't true that for all *x*, there exists a *y* such that *x* is less than *y*," or, more informally, "The counting numbers aren't infinite."

Suppose I want to test to see if this statement is a tautology. It better *not* be a tautology, because I already have an interpretation that makes it false.

As I point out in Chapter 8, to use a truth tree to see whether a statement is a tautology, negate that statement and use it as your trunk. When you complete the tree, if at least one open branch remains, the statement is a tautology; otherwise, it's either a contradiction or a contingent statement.

So, my first step is to negate the statement and use it as the trunk of my tree, and then decompose using **UI**, remembering *not* to check off the statement. To save space, I do both steps at once:

∀*x* ∃*y* [*Lxy*]

∃*y* [*Lxy*]

Next, I decompose this new statement using **EI**:

$$\forall x \exists y\,[Lxy]$$
$$\exists y\,[Lxy]\ \checkmark$$
$$Lxz$$

In this case, I decomposed the statement changing the variable to z in the hopes that I might be able to get rid of this variable later by applying **UI** again to the first statement.

At this point, the tree isn't complete because a branch is open and a statement isn't checked. But, there's still a chance I can close off the branch with a clever use of **UI** applied to the first statement. And, because the variable z is now in the mix, I use this constant in my decomposition:

$$\forall x \exists y\,[Lxy]$$
$$\exists y\,[Lxy]\ \checkmark$$
$$\boxed{Lxz}$$
$$\exists y\,[Lzy]$$

Now, another application of **EI**:

$$\forall x \exists y\,[Lxy]$$
$$\exists y\,[Lxy]\ \checkmark$$
$$\boxed{Lxz}$$
$$\exists y\,[Lzy]\ \checkmark$$
$$\boxed{Lzw}$$

This time, I introduce the variable w, but it doesn't matter. In fact, there's nothing you can do to complete the tree, either by closing off the branch or checking off the first statement. What you have here is a non-terminating tree.

Because the tree is never finished, it tells you nothing about whether the statement you are testing is a tautology. Non-terminating trees are the reason that truth trees, which are always useful in SL, have a more limited function in QL.

Part V
Modern Developments in Logic

The 5th Wave
By Rich Tennant

"That may be a valid argument, but we still need to get that ball out of there."

Part V
Modern
developments
in Logic

In this part . . .

*L*ogic may have started with Aristotle, but it certainly didn't end with him. Part V brings you completely up to date, discussing logic in the 20th century and beyond.

In Chapter 20, you discover how logic is instrumental to the computer at the levels of both hardware and software. Chapter 21 presents a few examples of non-classical logic — which are forms of logic that start off with different sets of assumptions than those forms I discuss in the rest of this book. I show you some startling differences between what seems obvious and what is possible in logic. Finally, Chapter 22 examines how paradoxes challenge logic, and how questions of consistency and completeness in logic led to the most important mathematical discovery of the century.

Chapter 20

Computer Logic

• •

• •

The computer has been called the most important invention of the 20th century (well, except for maybe the auto-drip coffee maker). And what sets the computer apart from other inventions — for example, the airplane, the radio, the television, or the nuclear power generator — is its versatility.

When you think about it, most machines are just tools for doing repetitive work that humans don't like to do — and honestly, machines generally do their jobs better than humans could ever do them. From the can opener to the car wash, machines have long been built to mimic human movement and then improve upon it.

So, it makes sense that along the way, people began to wonder whether a machine could be built to take over some of the repetitive *mental* labor that humans must do day in and day out. To some extent, adding machines and cash registers were invented to do just that. But, these inventions were also limited in what they could do. Just as you can't expect a can opener to wash a car, you can't expect an adding machine to do long division, let alone differential calculus.

A few visionaries, however, saw the possibility that a single machine might be capable of performing an unlimited number of functions.

In this chapter, I show you the role that logic played in the design of the computer. I start out with the beginnings of the computer — with the work of Charles Babbage and Ada Lovelace. Then, I discuss how Alan Turing showed, in theory at least, that a computer can perform any calculation that a human can. Finally, I focus on the ways in which logic forms the underpinning of the computer at the levels of both hardware and software.

The Early Versions of Computers

Even though the work of building the first electronic computers began in the 1940s, the idea and design for them began more than a century earlier. The computer began as a crackpot idea that never really got off the ground, and it developed into one of the most important inventions in history.

Babbage designs the first computers

Charles Babbage (1791–1871) is credited as the inventor of the computer. Even though his two models — the difference engine and the later analytical engine — were both designed to be powered by mechanical energy rather than electrical energy, they were highly sophisticated machines and had much more in common with later computers than with other inventions of the day.

Babbage began working on the difference engine in the 1820s. Unfortunately, even though he completed its design, he never finished building the machine. Funding difficulties and Babbage's personality conflicts with others on the project are cited as the reason for the failure to finish. (But, I'm sure that even modern-day computer engineers can relate to these obstacles.)

It wasn't until 1991 when the first and only difference engine was built in accordance with Babbage's plans. The builders limited themselves to the technology that would have been available in Babbage's time. They found that the machine worked in accordance with its plans, performing complex mathematical calculations with perfect accuracy.

After Babbage abandoned his plan to build the difference engine, he took up a more ambitious project — designing the analytical engine. This project incorporated the skills he had learned from the difference engine, but he took them a step further. One major improvement was that the analytical engine could be programmed with punch cards, making it more versatile and easier to use than the difference engine. Ada Lovelace, a mathematician and friend of Babbage's, assisted greatly in the design of the analytical engine. She also wrote several programs that would have run on the engine had it been built.

Turing and his UTM

After Charles Babbage's death in 1871, his designs gathered dust for decades until another visionary — Alan Turing (1912–1954) — approached the idea of mechanized computing from a different angle.

Turing saw the need to clarify exactly what was meant by computation, which humans had been doing for centuries using *algorithms*. Algorithms are simply mechanical procedures that produce a desired result in a finite number of steps. For example, the procedure for multiplying two numbers is an algorithm. As long as you follow the steps correctly, you're guaranteed to eventually get the right answer, no matter how big the numbers.

Turing found that algorithms could be broken down into small enough steps that they could be performed by an extremely simple machine. He called this machine the universal Turing machine (UTM).

Unlike Babbage, Turing never expected his machine to be built. Instead, it was a theoretical model that could be described in principle without being realized. However, the basic capability of a universal Turing machine is shared by all computers. In other words, every computer, no matter what its design, is neither more nor less capable of calculation than any other.

Explaining the UTM

The UTM consists of a paper strip of arbitrary length that's divided into boxes. Each box contains a single symbol from a finite set of symbols. The paper strip is on rollers so that it moves one box at a time past a pointer, which can both read the symbol in that square and, in some cases, erase it and write a different symbol.

For example, suppose you wanted to write a program to multiply two numbers together. You would begin by writing two numbers onto the strip, separating them with the multiplication operator:

∨
| | | | | | | | | | |7|5|8|X|6|3| | | | | | | | | | | | | |

These numbers would be the program's *initial conditions.* By the end of the procedure, the result would look like this:

∨
| | | | | | | | | | |7|5|8|X|6|3|=|4|7|7|5|4| | | | | | |

Turing laid out a set of allowable steps for getting from start to finish. These steps form what is called a *program.* A program consists of a list of *states* that tell the machine what *actions* to take based on what the pointer reads. Depending upon which state the machine is in, it performs different specific actions. The general form of these actions are as follows:

1. Optionally change the symbol in that box to a different symbol.

2. Optionally change the state that the machine is in.

3. Move one box to the right or left.

For example, when the machine is in state #1, and the pointer reads the number 7, the actions might be

1. Leave the 7 unchanged.

2. Change the machine from state #1 to state #10.

3. Move one box to the right.

However, when the machine is in state #2, and the pointer reads the number 7, the actions might be

1. Change the 7 to an 8.

2. Leave the state unchanged.

3. Move one box to the left.

Even though these actions seem crude, Turing showed that it's possible to perform sophisticated calculations in this way. Far more importantly, he proved that it's possible to perform *any* calculation in this way. That is, any algorithmic method that a human can learn to do, the UTM can also be programmed to do.

Relating the UTM to logic

But, you may ask, what does the UTM have to do with logic? To understand the connection, notice that the specifications for the machine make deciding the truth value of certain types of statements simple. For example, at the start of the multiplication example, this statement is true:

The machine is in state #1, and the pointer is on 7.

Based on the truth of this statement, the machine performs the proper actions, which resets the machine so that now this statement is true:

The machine is in state #10, and the pointer is on 5.

Logic is ideal for describing the conditions of the machine at any time. As you see in the next section, computer scientists have capitalized on the descriptive power of logic in the design of both hardware and software.

The Modern Age of Computers

The early ideas and theories Babbage and Turing — computer development pioneers — set the stage for the development of the modern computer. And, with electric power becoming more and more commonplace, the next advance in computers followed quickly. Built in the 1940s, ENIAC (short for Electronic Numerical Integrator and Computer) was the first electronic computer. Improved models developed swiftly.

All computers that have ever been developed (so far), including your home computer or laptop, can be divided into two levels of functionality:

- ✔ **Hardware:** The physical structure of the computer
- ✔ **Software:** The programs that run on the computer

As I explain in this section, logic plays an integral role in computing at both the hardware and software levels.

Hardware and logic gates

A *truth function* takes one or two input values and then outputs a value. Input and output values can be either of the two truth values (**T** or **F**). (See Chapter 13 for a refresher on truth functions.)

All five logical operators in sentential logic (SL) are truth functions. Three of them — the ~-operator, the &-operator, and the ∨-operator — are sufficient to build every possible truth function (see Chapter 13). Similarly, the |-operator (also known as Sheffer's stroke) by itself is also sufficient for building every possible truth function.

At the most basic level, computer circuits mimic truth functions. Instead of using the values **T** and **F**, however, computers use the values 1 and 0 as is done in Boolean algebra (see Chapter 14). In honor of George Boole, variables taking only these two values are called *Boolean variables*. When the current flow in a specific part of the circuit is on, the value there is said to be 1; when it's off, the value there is 0.

Types of logic gates

The behavior of operators is mimicked by *logic gates,* which allow currents to pass through circuits in a predetermined way. Six common gates are NOT, AND, OR, NAND, NOR, and XOR.

For example, here's a diagram for a NOT gate:

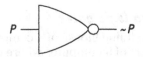

In Boolean terms, a NOT gate changes an input of 0 to an output of 1, and an input of 1 to an output of 0. In other words, when the current flowing into the gate is on, the current flowing out of it is off; when the current flowing in is off, the current flowing out is on.

The NOT gate is the only gate that has just one input. This single input parallels the ~-operator in that it's the only unary operator in SL. (By *unary*, I mean that the ~-operator is placed in front of *one* constant rather than placed between *two* constants.) The remaining gates all have two inputs. For example, take a look at a diagrams for an AND gate and an OR gate, respectively:

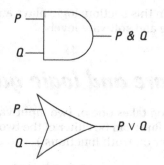

An AND gate turns its output current on only when both of its sources of input current are on; otherwise, its output current remains off. And, as you may have guessed, an OR gate outputs current when either of its input sources are on; otherwise, its output remains off.

Even though the NOT, AND, and OR gates are sufficient for imitating any possible SL truth function, for practical reasons having a greater variety of building blocks is helpful. The following three gates are also commonly used:

- **NAND:** Short for *not and,* this gate behaves in a similar fashion to the |-operator (see Chapter 13). Its output is 0 only when both of its inputs are 1; otherwise, its output is 0.

- **XOR:** Pronounced "ex-or," and short for *exclusive or,* this gate behaves like the exclusive *or* function that I discuss in Chapter 13. That is, when exactly one of its two inputs is 1, its output is also 1; otherwise, its output is 0.

- **NOR:** Short for *not or,* this gate turns on its output current only when both of its inputs are off; otherwise, its output current remains off.

Computers and logic gates

Gates are the building blocks of a computer's CPU (computer processing unit), which is the part of a computer where data is manipulated. Data is stored in memory devices and moved into and out of the CPU as needed, but the real "thinking" that goes on in a computer happens here in the CPU.

As Turing's work showed (see the earlier section "Turing and his UTM"), a short list of simple data manipulations is sufficient to calculate everything that can possibly be calculated. And logic gates provide more than enough of the basic functionality for reproducing a universal Turing machine.

Believe it or not, Charles Babbage's analytical engine, discussed earlier in this chapter, which ran programs on punch cards, would have no more and no less computing capability than today's state-of-the-art computers. Of course, today's computers have tons more memory and perform all processes with lightning speed compared with the Babbage model. But, in principle, both machines are sufficient to perform every possible mathematical calculation.

Software and computer languages

Computer hardware alone is already far more complex than just about any machine you can name. Still, no matter how complex the basic machinery of a computer is, it's still limited to perform whatever functions it was designed to perform — just like any other machine.

For some applications, these limited capabilities are all that are needed. Computer circuits for cars, watches, and household appliances, for example, do a fine job of regulating the machines they're built into. But, if you tried to use the circuit board from a BMW to run your dishwasher — or even a different type of car — you'd be disappointed at the results.

So why is your Mac or PC able to perform a seemingly endless number of tasks? The answer, of course, is software.

Software is any computer program that tells the hardware what tasks to carry out. All software is written in one of many *computer languages,* such as Java, C++, Visual Basic, COBOL, and Ada (named in honor of Ada Lovelace, whose work I mentioned earlier in this chapter). Even though all computer languages have differences in syntax and have various strengths and weaknesses, they all have one thing in common: logic.

As with SL, computer languages allow you to declare variables and constants to which you can then assign values. For example, suppose you want to declare a variable called "month" and set its initial value to 10. Here are the lines of computer code you would write in several different languages:

- **Java:** `int month = 10;`
- **Visual Basic:** `Dim Month as Integer = 10`
- **PL/I:** `DCL MONTH FIXED BINARY(31, 0) INIT(10);`

As you can see, in each language the syntax is different, but the main idea is the same. After you have variables to work with, you can build statements that test whether certain conditions are true or false — and then you can take action based on the results.

For example, suppose you want to send a message if and only if the date is October 10th. Here's how you might write an *if*-statement in each of the three languages to test whether the month and day are correct, and, if so, how to perform the proper action:

✔ **Java:**

```
if (month == 10 && day == 9)
        message = "Happy Leif Ericson Day!";
```

✔ **Visual Basic:**

```
If Month = 10 And Day = 9 Then _
        Message = "Happy Leif Ericson Day!"
```

✔ **PL/I:**

```
IF MONTH = 10 & DAY = 9 THEN
        MESSAGE = 'Happy Leif Ericson Day!';
```

Again, the syntax differs from language to language, but the meaning is essentially the same.

You can see how the structure of a programming language allows you to use a computer to do, in a more streamlined fashion, what a universal Turing machine can only do in very tiny steps. Nevertheless, the basic idea behind both types of computers is the same:

1. Set initial conditions.

2. Test the current conditions and make appropriate changes as needed.

3. Repeat Step 2 as often as necessary until complete.

Chapter 21

Sporting Propositions: Non-Classical Logic

. . . .

. . . .

To most people, the question "What if 2 + 2 = 5?" is absurd. I mean, c'mon, asking yourself "What if . . .?" in many cases is pointless because you know it could never be true. For example, consider this wacky question: "What if little green Martians landed their flying saucer on my front lawn and drove off in my car?"

But, to logicians, these "What if?" questions are a sporting proposition — a challenge that's inviting just because it's so preposterous. By the early 20th century, logic had been reduced to a relatively short list of basic assumptions called *axioms*. (See Chapter 22 for more on the axioms of logic.) These axioms were considered self-evident, and if you accepted them, everything else from logic followed.

But, what if you didn't accept them? What if, for the sake of argument, you changed an axiom the way that a baker changes an ingredient in a cake recipe?

Of course, a baker must choose carefully when changing ingredients. For example, replacing baking powder with sugar would stop a cake from rising, and replacing chocolate with garlic would result in a cake that even your dog wouldn't eat.

Similarly, a logician needs to choose carefully when changing axioms. Even a small change can result in a logical system that is full of contradictions. And even if a system were consistent, it may prove trivial, uninteresting, or useless — the garlic cake of logic. But given these caveats, a few mavericks have still fashioned alternative systems of logic.

Welcome to the world of *non-classical logic,* the modern response to more than 2,000 years of *classical logic* (which includes just about everything I discuss in this book). In this chapter, I give you a taste of several non-classical systems of logic. These systems may not only challenge your notions of true and false, but they may also surprise you in how useful they turn out to be in the real world.

Opening Up to the Possibility

In Chapter 1, I introduce you to the *law of the excluded middle* — which is the rule that a statement is either true or false, with no in-between allowed. This rule is an important assumption of logic, and for thousands of years it has served logic well.

When I introduce this law, I make it clear that not *everything* in the world fits together so neatly. Nor does it have to. The statement "Harry is tall" may be true when Harry is standing with a group of children, but it's false when he's standing next to the Boston Celtics.

Typically, logicians acknowledge this black-and-whiteness to be a limitation of logic. Logic doesn't handle shades of gray, so if you want to work with this statement, you must agree on a definition for the word *tall.* For example, you may define a man who stands six feet or more as *tall,* and all others as *not tall.*

The law of the excluded middle was a cornerstone of logic from Aristotle well into the 20th century. This law was one of those things you just accepted as a necessary assumption if you wanted to get anywhere with logic. But, in 1917, a man named Jan Lukasiewicz began wondering what would happen if he added a third value to logic. Check out the upcoming sections to see where Lukasiewicz's ponderings took him — and logic as a whole.

Three-valued logic

Jan Lukasiewicz decided to see what would happen if a third value, one that was neither true nor false, was added to logic. He called this value *possible,* and he stated that it could be assigned to statements whose truths were inconclusive, such as:

It will rain in Brooklyn tomorrow.

Doctors will someday find a cure for the common cold.

Life exists on other planets.

Possible seems to exist somewhere between true and false. For this reason, Lukasiewicz started out with the Boolean representation of 1 for true and 0 for false. (See Chapter 14 for more about Boolean algebra.) He then added a third value, ½, for the possible value.

The result was *three-valued logic*. Three-valued logic uses the same operators as classical logic, but with new definitions to cover the new value. For example, here's a truth table for ~x:

x	~x
1	0
½	½
0	1

This value for possibility makes sense when you use it in an example:

Let P = It will rain in Brooklyn tomorrow.

If the truth value P is ½ — that is, if it's possible — it's also possible that it won't rain in Brooklyn tomorrow. So, the value of ~P is also ½.

By a similar line of reasoning, you can figure out the truth value of &-statements and ∨-statements that include one or more sub-statements whose values are possible. For example:

Either George Washington was the first president *or* it will rain in Brooklyn tomorrow.

This statement is true (the first part is true, so the rest doesn't matter), so its value is 1.

Multi-valued logic

After the door is open for an intermediate truth value, you can create a logical system with more than one intermediate value — in fact, any number you choose. For example, imagine a system with 11 values, from 0 to 1 with increments of 1/10 in between. This system is an example of *multi-valued logic*.

0	¹⁄₁₀	²⁄₁₀	³⁄₁₀	⁴⁄₁₀	⁵⁄₁₀	⁶⁄₁₀	⁷⁄₁₀	⁸⁄₁₀	⁹⁄₁₀	1

How do you compute the truth value of statements with this system? In multi-valued logic, the rules are simple:

- **~-rule:** $\sim x$ means $1 - x$.
- **&-rule:** $x \& y$ means *choose the smaller value*.
- **∨-rule:** $x \lor y$ means *choose the larger value*.

These three rules may seem odd, but they work. And, you can also use them with three-valued logic and Boolean algebra.

For example:

$\sim 2/10 = 8/10$	**~-rule**
$3/10 \& 8/10 = 3/10$	**&-rule**
$1/10 \lor 6/10 = 6/10$	∨-**rule**
$7/10 \& 7/10 = 7/10$	**&-rule**
$9/10 \lor 9/10 = 9/10$	∨-**rule**

You can even add in the other two sentential logic (SL) operators using the logical equivalence rules **Impl** and **Equiv** (see Chapter 10 for more on equivalence rules):

- **Impl:** $x \rightarrow y = \sim x \lor y$
- **Equiv:** $x \leftrightarrow y = (x \rightarrow y) \& (y \rightarrow x)$

With these multi-value logic rules, you can calculate the value of any expression step by step, much as you would write a proof. For example:

$4/10 \leftrightarrow 3/10$	
$= (4/10 \rightarrow 3/10) \& (3/10 \rightarrow 4/10)$	**Equiv**
$= (\sim 4/10 \lor 3/10) \& (\sim 3/10 \lor 4/10)$	**Impl**
$= (6/10 \lor 3/10) \& (7/10 \lor 4/10)$	**~-rule**
$= 6/10 \& 7/10$	∨-**rule**
$= 6/10$	**&-rule**

Jan Lukasiewicz focused on the syntax of the system while leaving the semantics open to interpretation (see Chapter 14 for details on syntax and semantics). In other words, calculation in multi-valued logic is completely rule based, but the meaning of your result is open to interpretation.

(I describe several ways you might interpret results in the following section, "Fuzzy logic.")

Fuzzy logic

Some critics of multi-valued logic have questioned its usefulness to describe possible future events. They cite probability theory as the more appropriate tool for looking into the future. (Probability theory also calculates possibility, but it uses a different method of calculation.)

This criticism has some merit. For example, if the chance of rain in Brooklyn is 3/10 and in Dallas 7/10, probability theory calculates the chance of rain in *both* places by multiplying these values:

$$3/10 \times 7/10 = 21/100$$

On the other hand, multi-valued logic would calculate this value as 3/10. But, because probability theory is very well established, this discrepancy calls into question the usefulness of multi-valued logic.

In the 1960s, however, mathematician Lotfi Zadeh saw the potential for multi-valued logic to represent not what's possibly true but what's *partially true*.

Consider the following statement:

I am hungry.

After you eat a big meal, this statement will most likely be false. If you don't eat for several hours, though, eventually it will be true again. Most people, however, don't experience this change as black-and-white, but rather as shades of gray. That is, they find that the statement becomes more true (or false) over time.

In fact, most so-called opposites — tall or short, hot or cold, happy or sad, naïve or worldly, and so forth — aren't separated by a sharp line. Instead, they're ends of a continuum that connects subtle shadings. In most cases, shadings of these types are somewhat subjective.

Zadeh's answer to these issues is *fuzzy logic,* which is an extension of multi-valued logic. As with multi-valued logic, fuzzy logic permits values of 0 for completely false, values of 1 for completely true, or values in-between to describe shadings of truth. But, unlike multi-valued logic, all values between 0 and 1 are permitted.

Pricing out a new TV

To understand fuzzy logic a bit better, consider a hypothetical couple Donna and Jake. They have agreed to buy a brand-new television for their living room. Donna is thinking about a medium-sized model for around $500, but Jake has in mind a giant plasma screen that costs about $2,000. Of course, when they arrive at the store, the rude awakening occurs. If they look at their disagreement from the perspective of two-valued logic, they'll never make a decision.

But, after they start hashing things out, it's clear that some flexibility exists in both of their ideas.

As in multi-valued logic, &-statements in fuzzy logic are handled as the *smaller* of two values, and ∨-statements are handled as the *larger* of two values. In this case, the question calls for an &-statement: Both Donna *and* Jake must agree on an amount to spend.

A small amount of notation will be helpful:

Let $D(x)$ = Donna's truth value for x dollars.

Let $J(x)$ = Jake's truth value for x dollars.

Now you can set up this statement to capture their combined reaction to a particular price:

$D(x)$ & $J(x)$

So, when a salesman shows them televisions in the $500 range, he gets a 1 from Donna and a 0 from Jake:

$D(500)$ & $J(500)$ = 1 & 0 = 0

Then, when he shows the couple televisions in the $2,000 range, he gets a 0 from Donna and a 1 from Jake:

$D(2000)$ & $J(2000)$ = 0 & 1 = 0

When he finds the middle range, the impasse begins to loosen. Donna and Jake settle on a $1,100 projector television, which they both find to be the highest value possible:

$D(1100)$ & $J(1100)$ = .75 & .75 = .75

Buying votes

Of course, not all problems can be settled by &-statements. In some cases, you may find that a ∨-statement is the ticket. If so, fuzzy logic can still handle it.

For example, suppose you're a politician who needs one more vote from your city council to push through a land-use bill. The two abstaining voters are Rita and Jorge, both of whom are waiting to see how much money you allocate for a school playground project before casting their votes. Rita hopes you'll spend around $5,000, and Jorge wants you to spend about $25,000:

Because you only need one vote, you set up a ∨-statement as follows:

$R(x) \lor J(x)$

At first, you think a compromise might work, so you try to split the difference at $15,000:

$R(15000) \lor J(15000) = .1 \lor .1 = .1$

Not a very good outcome. With this scenario, both council members will be so unhappy that they'll most likely both vote against your bill. But then you try both $5,000 and $25,000:

$R(5000) \lor J(5000) = 1 \lor 0 = 1$

$R(25000) \lor J(25000) = 0 \lor 1 = 1$

What a difference! You can choose either of these values and make at least one council member happy, which is all you need to ensure victory.

Getting into a New Modality

Like multi-valued logic, *modal logic* attempts to handle not just the true and the false, but the possible. For this reason, modal logic introduces two new operators: the *possibility operator* and the *necessity operator.*

$\Diamond x$ = It is possible that x.

$\Box x$ = It is necessary that x.

For example:

Let C = Circles are round.

Then $\Diamond C$ means "It is possible that circles are round," and $\Box C$ means "It is necessary that circles are round."

The two modal operators are logically connected as follows:

$\Box x = \sim \Diamond \sim x$

For example, the statement "It is necessary that circles are round," is equivalent to the statement "It is not possible that circles are not round." Similarly, the following equation also holds true:

$$\Diamond x = \sim\Box\sim x$$

For example, the statement "It is possible that ghosts are real," is equivalent to the statement "It is not necessary that ghosts are not real."

One important aspect of modal logic is the distinction between *necessary truth* and *contingent truth*. To understand the distinction, consider the following example:

> Let S = It snowed yesterday in Brooklyn.

Suppose that it really did snow yesterday in Brooklyn. In that case, both of these statements are true:

S	TRUE
S ∨ ~S	TRUE

Even though both statements are true, a difference exists between the types of truth. The first statement is *contingently true* because its truth is contingent on what actually happened. That is, the situation might have been different. It's even possible that the weather report was inaccurate.

The second statement, however, is *necessarily true*. That is, regardless of the circumstances in Brooklyn yesterday, there is a logical necessity that the statement is true.

The shading of difference becomes clearer after I add the necessity operator:

$\Box S$	FALSE
$\Box S$ ∨ ~S	FALSE

In this case, the first statement says: "It is necessary that it snowed yesterday in Brooklyn." In modal logic, this statement is false, because a scenario could exist in which it didn't snow there yesterday. On the other hand, the second statement says: "It is necessary that it either snowed or didn't snow yesterday in Brooklyn." This statement is true, underscoring the stronger level of truth independent of real-world events.

TECHNICAL STUFF

Handling statements of indirect discourse

Modal logic is just one type of logic that attempts to include statements of *indirect discourse* that can't be translated into SL and QL. *Deontic logic* allows you to handle statements of obligation and permission. For example:

You *must* stop at a red light.

You *are allowed to* go through a yellow light.

Similarly, *epistemic logic* includes operators for handling statements of knowledge and belief. For example:

Arnie *knows that* Beth is waiting for him.

Arnie *believes that* Beth is waiting for him.

Taking Logic to a Higher Order

Recall that quantifier logic (QL) includes individual constants, individual variables, and property constants, but not property variables. This allows you to quantify individuals but not properties. For example, you can represent the sentence

All bankers are rich.

as

$\forall x \, [Bx \rightarrow Rx]$

To be specific, this statement tells you that anything that has the property of being a banker also has the property of being rich.

After you start focusing on properties themselves, however, you may want to discuss them in a logical fashion. For example, suppose you're hiring an assistant and you're looking for someone who's friendly, intelligent, and upbeat. In QL, it's easy to declare constants for these properties:

Let $F = x$ is friendly.

Let $I = x$ is intelligent.

Let $U = x$ is upbeat.

Then you can represent the following sentence:

> Nadia is intelligent and Jason is upbeat.

as

> *In & Uj*

However, with QL, you can't represent this statement:

> Nadia has every property that Jason has.

Second-order logic (also called *second-order predicate logic*) allows you to handle statements about properties that individuals possess by quantifying property variables rather than just individual variables. For example, suppose that X stands for some property. Having that variable set in place, you can now represent the previous statement as:

$$\forall X\,[Xj \rightarrow Xn]$$

The literal translation of this statement is "For every property X, if Jason has the property X, then Nadia has the property X."

You can also quantify individual variables. For example, consider the following statement:

> Someone has every property that Jason has.

You can represent this statement as:

$$\forall X\,\exists y\,[Xj \rightarrow Xy]$$

The literal translation for this representation is "For every property X, there exists an individual y such that if Jason has the property X, then y has the property X."

Moving Beyond Consistency

In a sense, *paraconsistent logic* is the flip side of multi-valued logic. In multi-valued logic, a statement may be *neither true nor false*. In paraconsistent logic, a statement may be *both true and false*. That is, every statement has at least one but possibly two truth values: **T** and/or **F**.

In other words, it's possible for a statement of the form

   $x \ \& \ {\sim}x$

to be true. To put it another way, paraconsistent logic allows you to break the law of non-contradiction by allowing a statement to be both true and false. (See Chapter 1 for more on the law of non-contradiction.)

This essential difference between paraconsistent and classical logic also creates further waves with one important rule of inference — disjunctive syllogism (**DS**). To refresh your memory, check out Chapter 9 and consider these constants:

   Let A = Ken lives in Albuquerque.

   Let S = Ken lives in Santa Fe.

In classical logic, **DS** allows you to make the following inference:

   $A \vee S, {\sim}A : S$

What this argument means is that if Ken lives in either Albuquerque or Santa Fe and he doesn't live in Albuquerque, then he lives in Santa Fe.

In paraconsistent logic, however, this inference doesn't hold true. Even though the failure of **DS** in paraconsistent logic may sound wacky at first, the reason makes sense when you think about it.

Suppose that both A and ${\sim}A$ are true, which is allowed in paraconsistent logic, no matter how odd it sounds. That is, Ken both *does* and *doesn't* live in Albuquerque. In that case, the statement

   $A \vee S$

is true even if S is false. And, as I just said, the statement

   ${\sim}A$

is also true. In this case, both of these statements are true, but the statement

   S

is false. In other words, the two premises are true and the conclusion is false, so **DS** is an invalid argument in paraconsistent logic.

In this way, paraconsistent logic remains consistent even though, by definition, it's inconsistent. Or, putting it another way, paraconsistent logic is both
consistent and inconsistent — which, when you think about it, is kind of the
whole idea in a nutshell. Or not.

Making a Quantum Leap

Have you ever seen a con artist play the shell game? He takes a pea, places it
on a table, and hides it under a walnut shell. Then he places two more empty
shells on the table next to it and deftly moves the three shells around on the
table. If you can guess which shell the pea is hidden under, you win.

This clever trick looks easy, but it has separated many a mark from his money.
Usually, the con artist is an expert in sleight of hand and can move the bean
into his palm and then under a different shell without being detected.

Subatomic particles also seem to operate in accordance with a cosmic shell
game that's easy to describe but very difficult to make sense of. These particles
are the building blocks that everything in the universe is made out of —
including you, me, and this book. And, the more that scientists find out about
how the universe works on this sub-microscopic level, the stranger it becomes.

Introducing quantum logic

One of the strangest aspects of the way the universe behaves at the very
smallest level is that logic itself breaks down. *Why* it happens is anybody's
guess, though a few theoretical physicists are hot on the trail of answers. But,
what happens is very well documented, and it's described by *quantum logic*.

As I fill you in on quantum logic, just remember two things:

- Quantum logic is science, not science fiction — countless scientific
  experiments have shown it to be the way things work.
- It just doesn't make sense.

So, as I describe it, don't worry that you're missing something that the rest of
the world understands. They don't understand *why* the universe works this
way any more than you will. So, just focus on *what* happens, and you'll be all
right.

Playing the shell game

Imagine a shell game with only two shells, played with subatomic particles, which I'll call *peas*, because the letter *p* is short for *particle*. In this realm, quantum logic runs the show. Now, begin with the understanding that the following statement is true:

> **Statement #1:** The pea is on the table and it's under either the left shell or the right shell.

In quantum logic, many of the basic assumptions of classical logic are left intact. A statement may have one of only two possible values: **T** or **F**. Similarly, the basic operators from SL work just the same here. So, in quantum logic, just as in SL, you can declare variables like these, as illustrated in Figure 21-1:

Let P = The pea is on the table.

Let Q = The pea is under the left shell.

Let R = The pea is under the right shell.

Figure 21-1:
A mind-bending shell game.

The pea is under one of these two shells.

The pea isn't under the left shell.

The pea isn't under the right shell.

With this structure in place, you can write a statement in English and translate it into symbols. For example, Statement #1 translated into quantum logic becomes this:

$$P \mathbin{\&} (Q \lor R)$$

So far, so good: Everything up until now has been the same as in SL. Now, if you were using SL, you could use the distributive law (see Chapter 10) to write an equivalent statement. But, in quantum logic, the distributive law doesn't apply, so you can't rewrite the previous statement as this:

$$(P \mathbin{\&} Q) \lor (P \mathbin{\&} R) \qquad \text{WRONG!}$$

But, before you toss this statement aside, notice that it translates back into English in this way:

Statement #2: Either the pea is on the table and it is under the left shell or the pea is on the table and it is under the right shell.

In other words, Statement #1 and Statement #2 aren't necessarily equivalent, nor does one necessarily imply the other. So, it's possible for Statement #1 to be true and Statement #2 to be false. That is, it's true that the pea is under one of the two shells, but it isn't under the left shell and it isn't under the right shell.

As you can see from this example, quantum logic fundamentally contradicts SL. It also contradicts everything that seems possible, normal, and sane. But that's the situation. The particles that make up the universe obey these laws. Spooky, huh?

Chapter 22

Paradox and Axiomatic Systems

• •

In This Chapter

▶ Understanding set theory and Russell's Paradox

▶ Seeing how the SL axiomatic system measures up

▶ Appreciating consistency and completeness

▶ Limiting mathematics with Gödel's Incompleteness Theorem

• •

Consider the following statement:

> This statement is false.

If the statement is true, then it must be false. However, if it's false, it must be true. This problem, called the *Liar's Paradox,* dates back to the Greeks.

At first glance, this paradox seems no more than a curiosity. But, in various forms, paradoxes like this one have surfaced and resurfaced, causing trouble for logicians and challenging them to search for ways to resolve them.

In this chapter, I explain Russell's Paradox (a modification of the Liar's Paradox), which forced logicians to make a radical restructuring of the foundations of set theory and logic. This leads to a discussion of the *Principia Mathematica,* which is an attempt to formulate set theory, logic, and ultimately all of math based upon a set of assumptions called axioms. You can also see how logic fares against the ultimate tests of mathematical certainty. And, finally, you can catch a glimpse of the limits of what can be proven logically as I introduce Gödel's incompleteness theorem.

Grounding Logic in Set Theory

Gottlob Frege's formulation of logic in the late 19th century depended on the relatively new work of Georg Cantor called *set theory.* Set theory provides a remarkably simple way to organize objects in the real world, but it also provides a unified way to define mathematical objects, such as numbers, by their mathematical properties.

In this section, I show you how set theory provided a natural foundation for logic, how this foundation was threatened, and how it was made solid again.

Setting things up

Set theory deals with, not surprisingly, sets. A *set* is simply a collection of things. For example:

> Let *S* = the set of all the shirts I own.

> Let *H* = the set of all the hats that you own.

A set is only properly defined when you can clearly distinguish what's *in* the set and what's *not*.

The items in a particular set are called *elements* of that set. For example, the shirt I'm wearing right now is an element of set *S*, and your favorite hat (assuming you own it) is an element of set *H*.

Sets may contain other sets, called *subsets*. For example:

> Let *B* = the set of all the blue shirts I own.

> Let *L* = the set of all the shirts I own that are in the laundry basket.

Both *B* and *L* are subsets of *S*. That is, any element that's in either of these sets is also in set *S*.

Even though set theory may seem rather simplistic, it's actually a powerful way to express logical ideas.

For example, consider the following statement

> All of my blue shirts are in the laundry basket.

This statement is easily expressible in quantifier logic (QL):

$$\forall x\,[Bx \rightarrow Lx]$$

This statement is true if and only if set *B* is a subset of set *L*:

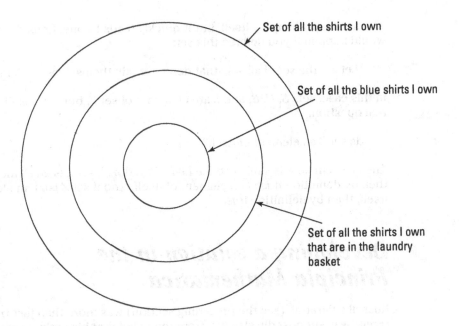

Set of all the shirts I own

Set of all the blue shirts I own

Set of all the shirts I own that are in the laundry basket

Though set theory and logic look different on the surface, they both express similar ideas. That's why Frege formally grounded his logic in the apparently solid ground of set theory. As long as set theory was free from contradictions, logic was presumed consistent as well. And set theory was so simple that it seemed unassailable.

Trouble in paradox: Recognizing the problem with set theory

It took a genius, Bertrand Russell, to spot a problem with set theory. In his honor, this flaw is called *Russell's Paradox*. This paradox hinges on the notion of *self-reference*. For example, the Liar's Paradox is a paradox of self-reference: The root of the problem is that the sentence speaks about itself.

Set theory faced a similar problem because it was possible for a set to contain itself as an element. Most of the time, however, sets don't contain themselves as elements. For instance, looking back to the shirt and hat example from the previous section, you see that set S contains only shirts and set H contains only hats. But, consider this set:

Let X = the set of all the sets mentioned in this chapter.

Set X contains sets S, H, B, and L as elements, of course, but it also contains *itself* as an element, because set X is mentioned in this chapter.

This isn't a problem by itself, but it quickly leads to one. Consider what would happen if you defined this set:

Let Z = the set of all sets that don't contain themselves as elements.

In this case, sets S, H, B, and L are elements of set Z, but set X isn't. Here's the real question:

Is set Z an element of itself?

The problem here is similar to the Liar's Paradox: If set Z is an element of itself, then by definition it *isn't* an element of itself. And if set Z isn't an element of itself, then by definition it *is*.

Developing a solution in the Principia Mathematica

Russell's Paradox (see the preceding section) was more than just troubling. (Frege was supposedly stricken from the belief that his entire work had been destroyed.) The paradox actually forced logicians to recast set theory and logic in a different way. The trick became figuring out how to keep the paradox from creeping in while still retaining the bulk of what was useful and descriptive in the original systems.

In an attempt to resolve the paradox, as well as to create a solid foundation for mathematics, in the first decade of the 20th century, Bertrand Russell and Alfred North Whitehead wrote the *Principia Mathematica*. This ambitious work was the first full-scale attempt to describe all of mathematics as a formal *axiomatic system* — an organization of mathematical ideas based on a small number of statements assumed to be true.

The core of an axiomatic system is a short list of simple statements called *axioms*. Axioms are combined in specifically defined ways to derive a much larger set of statements called *theorems*. Russell and Whitehead carefully chose their axioms with several purposes in mind:

- ✔ Creating a system powerful enough to derive sophisticated statements about mathematics as theorems.

- ✔ Avoiding *all* inconsistencies, such as Russell's Paradox.

- ✔ Showing that *all* possible mathematical truths could be derived as theorems.

Russell and Whitehead were certainly successful in achieving the first goal. And, their system also eliminated paradoxes of self-reference, such as Russell's Paradox. However, whether the Principia Mathematica could avoid *all* inconsistencies and provide a method to derive *all* of mathematics remained to be seen.

Even though the axiom set of the *Principia* did solve the problem of Russell's Paradox, in practice it was awkward, so it didn't catch on with mathematicians. Instead, a different set of axioms, the *Zermelo-Frankel axioms* (ZF axioms), solved the problem by distinguishing sets from more loosely defined objects known as *classes*.

Today, the words *set theory* usually refers to one of several versions of set theory based in the ZF axioms. On the other hand, simpler versions of set theory do exist that don't attempt to prevent such paradoxes. These versions are collectively called *naïve set theory*.

Understanding how axiomatic systems work is a major goal of this chapter. I discuss this concept in greater detail in the sections that follow.

Discovering the Axiomatic System for SL

To give you a sense of how theorems are formally derived in SL, this section shows you the basic structure of the axiomatic system that Russell and Whitehead developed for the *Principia Mathematica*.

In a nutshell, a formal axiomatic system has four requirements. All of these requirements are cast in terms of sets, because in logic and all other mathematics at the most formal level, *everything* is cast in terms of sets. That's why getting the bugs out of set theory, as I describe in the section "Grounding Logic in Set Theory," was so critical.

These four requirements are

- **Requirement #1:** A set of symbols
- **Requirement #2:** A set of rules for deciding which strings of symbols are well-formed formulas (WFFs)
- **Requirement #3:** A set of axioms
- **Requirement #4:** A set of rules for combining axioms and/or theorems to create new theorems

So, with those requirements in mind, you can see how well SL fits the definition of an axiomatic system. Requirement #1 is fulfilled because SL contains a set of symbols — operators, constants, and parentheses (see Chapter 4). Requirement #2 is fulfilled with the set of rules for WFFs (see Chapter 14).

To satisfy requirement #3, here are the four SL axioms from the *Principia Mathematica:*

1. $(x \lor x) \to x$

2. $x \to (x \lor y)$

3. $(x \lor y) \to (y \lor x)$

4. $(x \lor y) \to ((z \lor x) \to (z \lor y))$

As far as requirement #4, take a look at the following two rules for making new SL theorems:

- ✔ **Rule of Substitution:** In each case, you can substitute a constant or an entire statement for a variable, as long as you do so uniformly throughout.

   For example, two possible substitutions for axiom #1 are

   $(P \lor P) \to P$

   $((P \& Q) \lor (P \& Q)) \to (P \& Q)$

- ✔ **Modus Ponens (MP):** If you know that $x \to y$ and x are both theorems, you may add y to your list of theorems. See Chapter 9 for more on Modus Ponens.

From this short list of rules, it's possible to derive all of the rules of inference for SL, which I cover in Part III. This shows you that, even though this list of axioms is short, the theorems it allows you to derive are very powerful.

Proving Consistency and Completeness

With the formalization of SL as an axiomatic system came two important proofs about SL: In SL, every theorem is a tautology, and every tautology is a theorem. That is, theorems and tautologies are equivalent in SL.

Throughout this book you use both truth tables and proofs indiscriminately to draw conclusions about SL statements. After you know that every theorem is a tautology and vice versa, you have license to apply both syntactic methods (proofs) and semantic methods (truth tables) to a specific problem. (You can test whether a particular theorem is a tautology by simply making a truth table and checking it with the method I discuss in Chapter 6.)

Still, when you step back a moment, you should realize that the equivalence of theorems and tautologies in SL isn't to be taken for granted. Why isn't it the case, for example, that a theorem turns out not to be a tautology? Or, conversely, why isn't there a tautology that can't be produced as a theorem using the limited axioms and rules in the previous section?

The first question deals with the consistency of SL, and the second deals with the completeness of SL. I discuss both of these in the following sections.

Consistency and completeness of SL and QL

In 1921, mathematician Emil Post proved that SL is both consistent and complete. An axiomatic system is *consistent* if and only if every theorem generated in that system is a tautology. An axiomatic system is *complete* if and only if every tautology can be generated as a theorem in that system.

Although inconsistency loomed as a danger to axiomatic systems, threatening to undermine even the most carefully constructed work (for example, Gottlob Frege's logic), completeness was more like the Holy Grail. In other words, although inconsistency was something that mathematicians and logicians had to be careful of, completeness was just flat-out difficult to come by.

In fact, completeness was the unspoken goal of mathematics dating back to the Greeks. Euclid, for example, is honored as the founder of geometry even though geometry had been studied for hundreds or even thousands of years before him. (See Chapter 2 for more on Euclid.) Euclid's great insight was that geometry could be founded on five axioms, and that from these, all other true statements about geometry could be derived as theorems.

The decade following Emil Post's proof was a fruitful time for logic. In 1928, David Hilbert and William Ackerman proved that quantifier logic (QL) is consistent. Then, in 1931, Kurt Gödel proved that QL is complete. Moreover, this important result was his doctoral dissertation, which was the first work of the man who would become one of the greatest mathematicians of the 20th century.

Formalizing logic and mathematics with the Hilbert Program

By the 1920s, logic and mathematics had developed sufficiently for a precise examination of whether every mathematical truth could be shown as a theorem. Mathematician David Hilbert was instrumental in advocating the complete formalization of logic and mathematics. This formalization became known as the *Hilbert Program*.

Hilbert realized that what philosophers and mathematicians had been intuitively reaching for since the days of Aristotle and Euclid was now potentially in reach: a single axiomatic system, purged of inconsistencies, for expressing and calculating all logical and mathematical truths.

Hilbert's Program stressed the need for placing all of mathematics in strict axiomatic terms. All logical assumptions must be stated explicitly in formal language that lacked all ambiguity. (The *Principia Mathematica* was an example of an attempt at such formalization, and Hilbert studied it in depth.) From this foundation, the intuitive notion of mathematical proof could itself be formalized, resulting in *proof theory*.

Peano's axioms

One important difference exists between the axioms of the *Principia Mathematica* and the axioms of logic: The *Principia* axioms were strong enough to derive mathematically sophisticated statements. (See the section "Developing a solution in the Principia Mathematica" for more on the *Principia*.)

The *Principia* axioms made it possible to derive five foundational axioms of number theory — the foundation of all higher mathematics — developed by mathematician Giuseppe Peano:

1. Zero is a number.

2. If a is a number, then the successor of a is a number.

3. Zero is not the successor of a number.

4. Two numbers whose successors are equal are themselves equal.

5. If a set S contains zero and the successor of every number, then every number is in S. (This is called the *Induction Axiom*.)

In 1931, Gödel showed that the *Principia Mathematica,* an axiomatic system powerful enough to derive Peano's axioms (and thus model mathematics), was doomed to be either inconsistent or incomplete. More generally, he showed that *any* axiomatic system powerful enough to model mathematics was similarly doomed.

Hilbert championed the ongoing search for consistency and completeness proofs for axiomatic systems. Comparing an axiomatic system to a ship at sea, you could say that consistency means the ship won't sink and that completeness means the ship will take you everywhere you want to go.

Gödel's Incompleteness Theorem

The proofs that QL was both consistent and complete made mathematicians optimistic that the Hilbert Program would succeed. Ironically, the man who proved the completeness of QL would soon after demonstrate that the Hilbert Program would never succeed.

In 1931, Kurt Gödel published his Incompleteness Theorem, which states that no axiomatic system can have all three of the following properties:

- ✔ **Consistency:** Every theorem of the system is a tautology in an area that the system intends to model.

- ✔ **Completeness:** Every tautology in an area that the system intends to model is a theorem of the system.

- ✔ **Sufficient power to model mathematics:** The system can be applied as a model for mathematics.

The importance of Gödel's theorem

This theorem is generally considered to be the most important mathematical result of the 20th century. It's astounding because it places limits on what mathematics can do — or, more precisely, places limits on the degree to which axiomatic systems can describe mathematics. And, in one coup de grace, Gödel's Incompleteness Theorem demonstrated that the goal of the Hilbert Program (see the preceding section) was unattainable.

Interestingly, Gödel proved his conjecture by adopting a strategy along the same lines as the Liar's Paradox and Russell's Paradox: namely, self-reference. But, instead of being confounded by paradox, Gödel made paradox work in his favor. His strategy hinged on the fact that any system expressive enough to model complex mathematics would also be expressive enough to model itself — an extremely tricky task.

How he did it

As a first step, Gödel showed how to assign a unique number, called the *Gödel number,* to every string of symbols in a system. This numbering allowed him to uniquely number not only random strings but also statements, theorems, and even entire arguments, valid or otherwise. For example, all of these strings could be represented:

$$4 + 9 = 13$$
$$4 + 9 = 1962$$
$$\forall x \,\exists y \,[x + y = 3]$$
$$=48<+33-=7=$$

Because Gödel numbering worked at the level of the string, literally nothing that could be expressed in the system escaped his numbering, including that last string, which is meaningless.

Next, he showed how to use the system to build special statements called *meta-statements,* which referred to other statements. For example:

The statement "4 + 9 = 13" is a theorem.

The statement "4 + 9 = 13" is not a theorem.

The statement "4 + 9 = 13" includes the symbol "3."

There exists a proof that the statement "4 + 9 = 13" is a theorem.

These meta-statements themselves were just statements, each with its own Gödel number. In this way, one meta-statement could refer to another meta-statement. A meta-statement could even refer to itself. For example:

"There does not exist a proof that this statement is a theorem."

The paradox in the previous statement is even more subtle than it appears because Gödel built the statement to guarantee that at the level of semantics, the statement is a tautology. (The math behind this guarantee is completely over the top, so just take my word for it.) But then consider the following:

✓ If the statement *can* be derived as a theorem, then it isn't a theorem, so the contradiction of the statement can also be derived as a theorem, making the system by definition *inconsistent*.

✓ If the statement *can't* be derived as a theorem, then the system is by definition *incomplete*.

If it sounds complicated, that's because it is. And I've only scratched the surface. Understanding *what* Gödel accomplished is fascinating, but understanding *how* he accomplished it is something that even most mathematicians have only attained in a more or less sketchy fashion.

Pondering the Meaning of It All

Since Gödel published his proof, which undermines the usefulness of axiomatic systems to express mathematical truth, opinions have been divided about its meaning on a philosophical level.

In a sense, Gödel's Incompleteness Theorem was a response to Leibniz's 250-year-old dream of finding a system of logic powerful enough to calculate questions of law, politics, and ethics. (See Chapter 2 for more about Leibniz.) And Gödel's response was a definitive "No! You just can't do that!" Given that logic is insufficient for framing a complete model of mathematics, it certainly seems unlikely that it will ever adequately provide the tools to resolve ethical questions by mere calculation.

You might think that Gödel's proof implies that the rational mind is limited in its ability to understand the universe. But, though the mind may have its limitations, Gödel's result doesn't prove that these limitations exist. The proof just discusses how axiomatic systems are limited in how well they can be used to model other types of phenomena. The mind, however, may possess far greater capacities than an axiomatic system or a Turing machine.

Another common, and probably hasty, reaction to Gödel's work is to assume that his proof implies a limit on artificial intelligence. After all, human intelligence has developed rather well in this universe. Why couldn't other forms of intelligence, even artificial ones, develop along similar lines?

Like other uncanny scientific results of the 20th century, such as Relativity Theory and Quantum Mechanics, Gödel's proof answers one set of questions only to open up a new set of questions that are far more compelling.

Part VI
The Part of Tens

In this part . . .

Who doesn't love a couple of good top ten lists? Well, I sure hope you do because in this part, you get three top tens that give you some fun info on logic — some of which might just help you pass your next exam!

Chapter 23 shows off my ten favorite quotes about logic, from thinkers and cynics throughout the ages. In Chapter 24, I give you my nominees for the ten best logicians. Chapter 25 contains ten great tips for passing a logic exam.

Chapter 23

Ten Quotes about Logic

*O*kay, so I actually included 11 quotes here — but who's counting? They're all interesting and offer a variety of perspectives on logic.

"Logic is the beginning of wisdom, not the end."

Leonard Nimoy — American actor (speaking as Spock on *Star Trek*)

"Logic: The art of thinking and reasoning in strict accordance with the limitations and incapacities of the human misunderstanding."

Ambrose Bierce — American writer/satirist

"Logic is the anatomy of thought."

John Locke — 17th century English philosopher

"Pure logic is the ruin of the spirit."

Antoine de Saint-Exupery — French writer

"Logic is the art of going wrong with confidence."

Joseph Wood Krutch — American naturalist/writer

"Logic is like the sword — those who appeal to it shall perish by it."

Samuel Butler — English novelist/essayist

"You can only find truth with logic if you have already found truth without it."

G. K. Chesterton — English writer

"Logic will get you from A to B. Imagination will take you everywhere."

Albert Einstein — German physicist

"Logic takes care of itself; all we have to do is to look and see how it does it."

Ludwig Wittgenstein — Austrian philosopher

"Logic is in the eye of the logician."

Gloria Steinem — American activist/writer

"You can use logic to justify anything. That's its power and its flaw."

Kate Mulgrew — American actor
(speaking as Captain Kathryn Janeway on Star Trek *Voyager*)

Chapter 24

Ten Big Names in Logic

In This Chapter

▶ Seeing how Aristotle changed it all

▶ Discovering the people who transformed logic

▶ Understanding the evolution of logic

Here are my nominees for the Logic Hall of Fame. Lots of great minds had to be passed over, or this list would be miles long, but here are my top ten.

Aristotle (384–322 BC)

Aristotle was the originator of logic. Before him, philosophers (such as Socrates and Plato) and mathematicians (such as Pythagoras and Thales) presented arguments on a wide variety of topics. But, Aristotle was the first to examine the structure of argument itself.

In a series of six philosophical writings on logic, later collected as a single work titled *Organon,* Aristotle identified the foundational concepts in logic. He defined a statement as a sentence that possesses either truth or falsehood (for more on statements, flip to Chapters 1 and 3). He also studied valid argument structures called syllogisms (see Chapter 2), which contained premises that led inevitably to a conclusion. For centuries after his death, Aristotle's writings on logic were often studied and commented upon, but rarely surpassed. (See Chapter 2 for more fun facts about Aristotle.)

Gottfried Leibniz (1646–1716)

A bona fide Renaissance man, Gottfried Leibniz was the first philosopher in the Age of Reason to see the potential for logic to be used as a tool for calculation. He hoped that logical calculation would someday be on par with mathematics. He even worked out the beginnings of a symbolic representation of logic, anticipating formal logic by 200 years. (Check out Chapter 2 to read more about Leibniz.)

George Boole (1815–1864)

George Boole invented Boolean algebra, which was the prototype for formal logic. Boolean algebra was the first system of logic that used pure calculation to determine the truth value of statements. Boole used 1 to represent true and 0 to represent false. Computer scientists still use Boolean variables as objects that can take only these two values and no others. (See Chapter 2 for more on Boole and Chapter 14 for a look at Boolean algebra.)

Lewis Carroll (1832–1898)

Even though he's most famous as the author of *Alice in Wonderland,* Lewis Carroll (whose real name was Charles Dodgson), a professor of mathematics at Cambridge University in England, also wrote several books on logic. He also delighted in writing logic puzzles. Here's an old favorite of his:

> Babies are illogical.
>
> Nobody is despised who can manage a crocodile.
>
> Illogical persons are despised.

The goal here is to use all three premises to arrive at a logical conclusion, which in this case is "Babies can't manage crocodiles."

To be fair, Carroll probably shouldn't be on this list — his contributions to logic were mostly recreational. But, then again, he was a logician and he certainly is a big name, so one might equally draw the conclusion that he's a big name in logic, logically speaking.

Georg Cantor (1845–1918)

Georg Cantor was the inventor of set theory, which was the foundation of logic and, arguably, for all other mathematics. (Flip to Chapter 2 for more on Cantor and Chapter 22 for more on set theory.) He was also the first to incorporate into math an understanding of infinity as a calculable entity instead of just a mysterious phantom. For all these achievements and more, Cantor is on everybody's short list for the greatest mathematician of the 19th century.

Gottlob Frege (1848–1925)

Gottlob Frege, inventor of formal logic, built upon the work of Boole, Cantor, and others to develop the first logical systems recognizable as what were to become sentential logic and quantifier logic. His logic included the five logical operators *not, and, or, if,* and *if and only if.* It also included symbols for *all* and *there exists.* Check out Chapter 2 to read more about Frege and his contribution to logic.

Bertrand Russell (1872–1970)

Bertrand Russell's nearly-100-year life span included such notable achievements as Russell's paradox and co-authorship of the *Principia Mathematica* with Alfred North Whitehead.

Russell's paradox resulted in a reformulation of Frege's logic and Cantor's set theory — both foundational systems that had previously appeared unassailable. The *Principia Mathematica* was Russell's attempt to formulate mathematics with perfect logical consistency and completeness. See Chapter 2 for further discussion of Russell and his place in the history of logic and mathematics.

David Hilbert (1862–1943)

David Hilbert was tremendously influential in both logic and mathematics. He advocated for the rigorous reduction of all mathematics to axioms (self-evident truths) and theorems (statements that could be logically proved from axioms). This trend in mathematics became known as the Hilbert Program, whose goal became the creation of an axiom system for mathematics that was both consistent and complete — that is, produced all possible true theorems and no false ones.

Although Kurt Gödel proved that the goal of the Hilbert Program wasn't attainable, Hilbert's contribution to the development of mathematical logic is undeniable. (See Chapter 22 for more on Hilbert.)

Kurt Gödel (1906–1978)

Gödel proved that logic, in its most powerful form (quantifier logic), is mathematically both *consistent* and *complete*. Consistency means that logic is free from contradictions. Completeness means that all true logical statements can be proved by syntactic methods. (See Chapter 7 for a discussion of syntactic proof.)

Gödel is most famous, however, for his proof that mathematics as a whole does *not* possess consistency and completeness. He said that any mathematical system that's free from contradictions must contain statements that are true but can't be proved as true using the axioms of that system. This discovery signaled the end of the Hilbert Program and is generally considered the most important mathematical result of the 20th century. See Chapters 2 and 22 for more details on Gödel's important work.

Alan Turing (1912–1954)

Alan Turing proved that all calculations that humans perform can be equally accomplished by a computer having a specified set of simple functions. These functions include the ability to check whether certain types of conditions are true or false and to take action based on the result.

Turing called any computer of this kind a universal Turing machine (UTM). Because every modern computer is a UTM, and logic is at the heart of how any UTM processes data, logic is a cornerstone of computer science. (Flip to Chapter 20 for more on how logic and computer science work together.)

Chapter 25

Ten Tips for Passing a Logic Exam

. . . .

In This Chapter

▶ Discovering techniques to help you effectively write a logic exam

▶ Figuring the quickest way out of a jam

▶ Understanding the importance of checking your work and admitting your mistakes

. . . .

So, let me guess — you're frantically reading this book eight hours before your big logic final? Not to worry. This quick chapter offers ten tips to help you excel at exam time. Read on — the grade you save may be your own!

Breathe

As you sit down and wait for your prof to hand out the exam, take a deep breath in to a slow count of five and then breathe out in the same way. Repeat. That's all. Do this for a minute or so (no more than that — you don't want to hyperventilate!). You'll find that you're now much calmer.

Start by Glancing over the Whole Exam

It takes only a minute, but looking through the entire test gets your brain working on the problems subconsciously as early as possible. As a result, a few of the problems may work themselves out more easily as you go along.

This technique also allows you to find helpful clues. For example, Question 3 may ask you to define a term that appears in Questions 5, 6, and 7.

Warm up with an Easy Problem First

Why shouldn't you warm up with an easy problem? Athletes know the importance of a warm-up to get everything moving. And, I guarantee that just putting something down on paper will make you feel better instantly.

Fill in Truth Tables Column by Column

When filling in a truth table, you can do it the easy way or the hard way. The easy way is column by column, as I show in Chapter 6.

If you try to go row by row instead, you'll have to keep switching gears, and it will take longer to achieve the same result.

If You Get Stuck, Jot Down Everything

Because proofs look so neat in the book, some students think they have to figure them out in their heads before they begin.

But writing proofs is a messy process, so feel free to be a mess. Use scratch paper to write down every possible path you can think of. After the big "Aha!" happens, start writing the proof neatly on the page you intend to turn in.

If You REALLY Get Stuck, Move On

It happens: The answer is staring you square in the face and you can't see it. The minutes are whizzing away, your heart is pounding, and your palms are so sweaty that your pencil is dripping.

If you're the praying kind, this would be a good time to start. But, remember, the Lord helps those who help themselves, and so I say unto thee: Moveth on!

Better to miss this one question than the five that you never got to because you got stuck. If you have time, you can always go back. And when you do, the answer may just jump out at you.

If Time Is Short, Finish the Tedious Stuff

Generally speaking, truth tables and truth trees tend to be plug-and-chug methods: Slow and steady work will always get you there. Quick tables and proofs, on the other hand, tend to be creative methods: They can go quickly if you have the right insight, but you have no guarantees.

So, when the professor calls out "Ten minutes left!" and she really means it, put aside the ornery proof that just isn't speaking to you at the moment and pick up the tedious truth table you've been avoiding for half the exam. As the exam grinds to its close, you're more likely to make progress if you work on the table instead of just staring at the impenetrable proof.

Check Your Work

I'm aware that in certain circles, checking your work on an exam is considered one of those quaint, spinsterly virtues — like carrying a monogrammed hanky or removing your hat upon entering a building.

But think about the last time you *didn't* check over your exam before handing it in. Did you really do something important with those seven or eight minutes you saved? No, you probably spent those precious minutes out in the hall with your friends from class, yapping about the exam.

And don't you just hate getting your exam back *after* it has been graded and catching those silly errors only because they're circled in red?

All right, so I'm sneezing into the wind here. But, even if you catch only one little three-point mistake, you may just bump yourself up a partial letter grade. And catching 5-, 10-, or even 20-point mistakes is not that uncommon.

Admit Your Mistakes

I know that this piece of advice flies in the face of all your survival instinct, but that's why I'm telling you.

For example, consider this tragic story: You've sweated for half an hour over a hairy proof trying to prove (P & Q). And as the professor starts calling for your blue books, you get to the last line and find you've proved ~(P & Q). Ouch! If you had time, you could go through your proof line by line and find the error, but time is exactly what you don't have.

Your first instinct — maybe the prof won't notice, and she'll give me full credit — is dead wrong. (Your second instinct, which is to quit school and move to a Tibetan monastery, is also probably not a good idea.) If you leave the mistake as is, your professor will think that you don't know the difference between a statement and its negation.

Instead, circle the error and write a note: "I know this is wrong, but no time to fix it." Now when the professor reads through it, she has some context. If you just made a minor mistake, such as dropping a ~ operator, she'll probably just ding you for a point or two. Even if you screwed up royally, at least she knows that you're aware of it. So, now she knows you're not completely clueless and you're rather honest as well. Either way, you win!

Stay Until the Bitter End

Keep writing until the professor pries the pencil from your blistered fingers. You're bound to pick up a few points in those last few minutes, and the prof will notice your extra effort.

Index

Russell, Bertrand
contributions of, 339
laws of thought, 15–17
Principia Mathematica, 30, 326–330
relationship to Frege's formal logic, 30
Russell's Paradox, relationship to set
theory, 325–326

• S •

science, using logic in, 47
scope
of operators, 79–80
of quantifiers in QL (quantifier logic), 236
second-order logic, explanation of, 318
self-referential statements, significance of,
283
semantic equivalence
checking with truth trees, 141–144
judging, 94–96
linking with tautology, 102–103
significance of, 207–208
strategic assumptions for, 114–115
testing with truth tables, 101
semantic inequivalence
checking with truth trees, 141–144
strategic assumptions for, 114–115
testing with truth tables, 101
semantically inequivalent statements,
connecting, 106
semantics
in Boolean algebra, 221–222
significance of, 213
semiring, explanation of, 221
sentential logic. *See* SL (sentential logic)
set theory
development by Georg Cantor, 338
elements in, 324
overview of, 27–28, 324–325
relationship to logic, 325
significance of, 327
subsets in, 324

sets
organizing concepts with, 12
relationship to intersections, 12
Sheffer's stroke, using, 210–211
shell game, relationship to quantum logic,
320–322
Simp & rule
breaking down premises with, 189
explanation of, 162
restriction related to QL statements, 255
using in proofs, 154–155, 173, 201
using with EG (Existential Generalization)
QL rule, 263
single branching statements
illustration of, 127
relationship to truth trees, 126
SL (sentential logic)
advantage of, 66
as axiomatic system, 328
axiomatic system for, 327–328
binary operators in, 64
versus Boolean algebra, 221–222
comparing to arithmetic, 63–65
comparing to Boolean algebra, 217–222
consistency and completeness of, 329
explanation of, 28, 52
identifying statements in, 214
implication rules of, 149–150
versus QL (quantifier logic), 226–227
representing basic statements in, 275
as symbolic language, 52
translating English to, 68–71
translating from English to, 242
translating implication rules into QL,
253–255
translating to English, 66–68
using constants in, 64–65
using parentheses (()) in, 76
using with valid arguments, 225
SL rules, applying in QL (quantifier logic),
252–256

strategic assumptions. *See also*
 assumptions
 beginning quick tables with, 110
 for consistency, 115
 for contingent statements, 114
 for contradictions, 114
 disproving, 112–113
 for inconsistency, 115
 for invalidity, 115
 for semantic equivalence, 114–115
 for semantic inequivalence, 114–115
 for tautologies, 113–114
 for validity, 115–116
strategy, planning for quick tables,
 113–116, 118
strings
 definition of, 215
 identifying as WFFs, 216–217
subsets
 role in sets, 324
 using in set theory, 28
sub-statements. *See also* statements
 building in proofs, 153
 identifying, 78
 including in if-statements, 11
substitution rule, applying in axiomatic
 systems, 328
syllogistic logic, Aristotle's invention of,
 20–23
symbolic logic. *See* formal logic
syntax
 in Boolean algebra, 221–222
 significance of, 213

• T •

T
 including in truth tables, 88
 using with formal logic, 53
T and F values, mixing in Boolean algebra,
 220
T in SL, corresponding symbol in Boolean
 algebra, 218

tables. *See* quick tables; truth tables
Taut (tautology) equivalence, using, 172
tautologies
 converting into contradictions, 101–102
 examples of, 41–42
 identification of, 85
 linking semantic equivalence with,
 102–103
 relationship to theorems, 328
 separating with truth trees, 134–137
 SL statements as, 93
 strategic assumptions for, 113–114
 testing for QL tree, 296–297
 testing with truth tables, 101
theorems
 in axiomatic systems, 328
 definition of, 23
 relationship to tautologies, 328
there exists, relationship to intersection, 12
there is no, relationship to intersection, 13
there is, relationship to intersection, 12
thinking versus logic, 40–41
though, similarity to *and*, 69
three-valued logic, overview of, 310–311
true statements, forms of, 126
trunk of truth tree
 constructing, 128
 constructing in QL, 288
truth function, explanation of, 305
truth in modal logic, types of, 316
truth tables. *See also* quick tables; truth
 trees
 capabilities of, 85, 87
 comparing to quick tables, 109
 constants and rows in, 88
 determining statement consistency with,
 96–98
 determining valid arguments with, 98–100
 filling in, 89–92, 342
 judging semantic equivalence with, 94–96
 for ?-operator, 206
 versus quick tables, 122
 reading, 55, 92–93

BUSINESS, CAREERS & PERSONAL FINANCE

0-7645-5307-0

0-7645-5331-3 *˜

Also available:

- Accounting For Dummies ˜
  0-7645-5314-3
- Business Plans Kit For Dummies ˜
  0-7645-5365-8
- Cover Letters For Dummies
  0-7645-5224-4
- Frugal Living For Dummies
  0-7645-5403-4
- Leadership For Dummies
  0-7645-5176-0
- Managing For Dummies
  0-7645-1771-6

- Marketing For Dummies
  0-7645-5600-2
- Personal Finance For Dummies *
  0-7645-2590-5
- Project Management For Dummies
  0-7645-5283-X
- Resumes For Dummies ˜
  0-7645-5471-9
- Selling For Dummies
  0-7645-5363-1
- Small Business Kit For Dummies *˜
  0-7645-5093-4

HOME & BUSINESS COMPUTER BASICS

0-7645-4074-2

0-7645-3758-X

Also available:

- ACT! 6 For Dummies
  0-7645-2645-6
- iLife '04 All-in-One Desk Reference
  For Dummies
  0-7645-7347-0
- iPAQ For Dummies
  0-7645-6769-1
- Mac OS X Panther Timesaving
  Techniques For Dummies
  0-7645-5812-9
- Macs For Dummies
  0-7645-5656-8

- Microsoft Money 2004 For Dummies
  0-7645-4195-1
- Office 2003 All-in-One Desk Reference
  For Dummies
  0-7645-3883-7
- Outlook 2003 For Dummies
  0-7645-3759-8
- PCs For Dummies
  0-7645-4074-2
- TiVo For Dummies
  0-7645-6923-6
- Upgrading and Fixing PCs For Dummies
  0-7645-1665-5
- Windows XP Timesaving Techniques
  For Dummies
  0-7645-3748-2

FOOD, HOME, GARDEN, HOBBIES, MUSIC & PETS

0-7645-5295-3

0-7645-5232-5

Also available:

- Bass Guitar For Dummies
  0-7645-2487-9
- Diabetes Cookbook For Dummies
  0-7645-5230-9
- Gardening For Dummies *
  0-7645-5130-2
- Guitar For Dummies
  0-7645-5106-X
- Holiday Decorating For Dummies
  0-7645-2570-0
- Home Improvement All-in-One
  For Dummies
  0-7645-5680-0

- Knitting For Dummies
  0-7645-5395-X
- Piano For Dummies
  0-7645-5105-1
- Puppies For Dummies
  0-7645-5255-4
- Scrapbooking For Dummies
  0-7645-7208-3
- Senior Dogs For Dummies
  0-7645-5818-8
- Singing For Dummies
  0-7645-2475-5
- 30-Minute Meals For Dummies
  0-7645-2589-1

INTERNET & DIGITAL MEDIA

0-7645-1664-7

0-7645-6924-4

Also available:

- 2005 Online Shopping Directory
  For Dummies
  0-7645-7495-7
- CD & DVD Recording For Dummies
  0-7645-5956-7
- eBay For Dummies
  0-7645-5654-1
- Fighting Spam For Dummies
  0-7645-5965-6
- Genealogy Online For Dummies
  0-7645-5964-8
- Google For Dummies
  0-7645-4420-9

- Home Recording For Musicians
  For Dummies
  0-7645-1634-5
- The Internet For Dummies
  0-7645-4173-0
- iPod & iTunes For Dummies
  0-7645-7772-7
- Preventing Identity Theft For Dummies
  0-7645-7336-5
- Pro Tools All-in-One Desk Reference
  For Dummies
  0-7645-5714-9
- Roxio Easy Media Creator For Dummies
  0-7645-7131-1

* Separate Canadian edition also available
˜ Separate U.K. edition also available

Available wherever books are sold. For more information or to order direct: U.S. customers visit www.dummies.com or call 1-877-762-2974.
U.K. customers visit www.wileyeurope.com or call 0800 243407. Canadian customers visit www.wiley.ca or call 1-800-567-4797.

WILEY

SPORTS, FITNESS, PARENTING, RELIGION & SPIRITUALITY

0-7645-5146-9

0-7645-5418-2

Also available:
- Adoption For Dummies
  0-7645-5488-3
- Basketball For Dummies
  0-7645-5248-1
- The Bible For Dummies
  0-7645-5296-1
- Buddhism For Dummies
  0-7645-5359-3
- Catholicism For Dummies
  0-7645-5391-7
- Hockey For Dummies
  0-7645-5228-7

- Judaism For Dummies
  0-7645-5299-6
- Martial Arts For Dummies
  0-7645-5358-5
- Pilates For Dummies
  0-7645-5397-6
- Religion For Dummies
  0-7645-5264-3
- Teaching Kids to Read For Dummies
  0-7645-4043-2
- Weight Training For Dummies
  0-7645-5168-X
- Yoga For Dummies
  0-7645-5117-5

TRAVEL

0-7645-5438-7

0-7645-5453-0

Also available:
- Alaska For Dummies
  0-7645-1761-9
- Arizona For Dummies
  0-7645-6938-4
- Cancún and the Yucatán For Dummies
  0-7645-2437-2
- Cruise Vacations For Dummies
  0-7645-6941-4
- Europe For Dummies
  0-7645-5456-5
- Ireland For Dummies
  0-7645-5455-7

- Las Vegas For Dummies
  0-7645-5448-4
- London For Dummies
  0-7645-4277-X
- New York City For Dummies
  0-7645-6945-7
- Paris For Dummies
  0-7645-5494-8
- RV Vacations For Dummies
  0-7645-5443-3
- Walt Disney World & Orlando For Dummies
  0-7645-6943-0

GRAPHICS, DESIGN & WEB DEVELOPMENT

0-7645-4345-8

0-7645-5589-8

Also available:
- Adobe Acrobat 6 PDF For Dummies
  0-7645-3760-1
- Building a Web Site For Dummies
  0-7645-7144-3
- Dreamweaver MX 2004 For Dummies
  0-7645-4342-3
- FrontPage 2003 For Dummies
  0-7645-3882-9
- HTML 4 For Dummies
  0-7645-1995-6
- Illustrator cs For Dummies
  0-7645-4084-X

- Macromedia Flash MX 2004 For Dummies
  0-7645-4358-X
- Photoshop 7 All-in-One Desk
  Reference For Dummies
  0-7645-1667-1
- Photoshop cs Timesaving Techniques
  For Dummies
  0-7645-6782-9
- PHP 5 For Dummies
  0-7645-4166-8
- PowerPoint 2003 For Dummies
  0-7645-3908-6
- QuarkXPress 6 For Dummies
  0-7645-2593-X

NETWORKING, SECURITY, PROGRAMMING & DATABASES

0-7645-6852-3

0-7645-5784-X

Also available:
- A+ Certification For Dummies
  0-7645-4187-0
- Access 2003 All-in-One Desk
  Reference For Dummies
  0-7645-3988-4
- Beginning Programming For Dummies
  0-7645-4997-9
- C For Dummies
  0-7645-7068-4
- Firewalls For Dummies
  0-7645-4048-3
- Home Networking For Dummies
  0-7645-42796

- Network Security For Dummies
  0-7645-1679-5
- Networking For Dummies
  0-7645-1677-9
- TCP/IP For Dummies
  0-7645-1760-0
- VBA For Dummies
  0-7645-3989-2
- Wireless All In-One Desk Reference
  For Dummies
  0-7645-7496-5
- Wireless Home Networking For Dummies
  0-7645-3910-8

HEALTH & SELF-HELP

0-7645-6820-5 *˙

0-7645-2566-2

Also available:
- Alzheimer's For Dummies
  0-7645-3899-3
- Asthma For Dummies
  0-7645-4233-8
- Controlling Cholesterol For Dummies
  0-7645-5440-9
- Depression For Dummies
  0-7645-3900-0
- Dieting For Dummies
  0-7645-4149-8
- Fertility For Dummies
  0-7645-2549-2

- Fibromyalgia For Dummies
  0-7645-5441-7
- Improving Your Memory For Dummies
  0-7645-5435-2
- Pregnancy For Dummies ˙
  0-7645-4483-7
- Quitting Smoking For Dummies
  0-7645-2629-4
- Relationships For Dummies
  0-7645-5384-4
- Thyroid For Dummies
  0-7645-5385-2

EDUCATION, HISTORY, REFERENCE & TEST PREPARATION

0-7645-5194-9

0-7645-4186-2

Also available:
- Algebra For Dummies
  0-7645-5325-9
- British History For Dummies
  0-7645-7021-8
- Calculus For Dummies
  0-7645-2498-4
- English Grammar For Dummies
  0-7645-5322-4
- Forensics For Dummies
  0-7645-5580-4
- The GMAT For Dummies
  0-7645-5251-1
- Inglés Para Dummies
  0-7645-5427-1

- Italian For Dummies
  0-7645-5196-5
- Latin For Dummies
  0-7645-5431-X
- Lewis & Clark For Dummies
  0-7645-2545-X
- Research Papers For Dummies
  0-7645-5426-3
- The SAT I For Dummies
  0-7645-7193-1
- Science Fair Projects For Dummies
  0-7645-5460-3
- U.S. History For Dummies
  0-7645-5249-X

Get smart @ dummies.com®

- **Find a full list of Dummies titles**
- **Look into loads of FREE on-site articles**
- **Sign up for FREE eTips e-mailed to you weekly**
- **See what other products carry the Dummies name**
- **Shop directly from the Dummies bookstore**
- **Enter to win new prizes every month!**

* Separate Canadian edition also available
˙ Separate U.K. edition also available

Available wherever books are sold. For more information or to order direct: U.S. customers visit www.dummies.com or call 1-877-762-2974.
U.K. customers visit www.wileyeurope.com or call 0800 243407. Canadian customers visit www.wiley.ca or call 1-800-567-4797.